Meeting the Nation's Needs for Biomedical *and* Behavioral Scientists

Committee on National Needs for
Biomedical and Behavioral Research Personnel

Studies and Surveys Unit

Office of Scientific and Engineering Personnel

National Research Council

NATIONAL ACADEMY PRESS
Washington, D.C. 1994

NATIONAL ACADEMY PRESS • 2101 Constitution Avenue, N.W. • Washington, DC 20418

NOTICE: The project that is the subject of this report was approved by the Governing Board of the National Research Council, whose members are drawn from the councils of the National Academy of Sciences, the National Academy of Engineering, and the Institute of Medicine. The members of the committee responsible for the report were chosen for their special competencies and with regard to appropriate balance.

This report has been reviewed by persons other than the author according to procedures approved by a Report Review Committee consisting of members of the National Academy of Sciences, the National Academy of Engineering, and the Institute of Medicine.

The National Academy of Sciences is a private, nonprofit, self-perpetuating society of distinguished scholars engaged in scientific and engineering research, dedicated to the furtherance of science and technology and to their use for the general welfare. Upon the authority of the charter granted to it by the Congress in 1863, the Academy has a mandate that requires it to advise the federal government on scientific and technical matters. Dr. Bruce Alberts is president of the National Academy of Sciences.

The National Academy of Engineering was established in 1964, under the charter of the National Academy of Sciences, as a parallel organization of outstanding engineers. It is autonomous in its administration and in the selection of its members, sharing with the National Academy of Sciences the responsibility for advising the federal government. The National Academy of Engineering also sponsors engineering programs aimed at meeting national needs, encourages education and research, and recognizes the superior achievements of engineers. Dr. Robert M. White is president of the National Academy of Engineering.

The Institute of Medicine was established in 1970 by the National Academy of Sciences to secure the services of eminent members of appropriate professions in the examination of policy matters pertaining to the health of the public. The Institute acts under the responsibility given to the National Academy of Sciences by its congressional charter to be an adviser to the federal government and, upon its own initiative, to identify issues of medical care, research, and education. Dr. Kenneth Shine is president of the Institute of Medicine.

The National Research Council was organized by the National Academy of Sciences in 1916 to associate the broad community of science and technology with the Academy's purposes of furthering knowledge and advising the federal government. Functioning in accordance with general policies determined by the Academy, the Council has become the principal operating agency of both the National Academy of Sciences and the National Academy of Engineering in providing services to the government, the public, and the scientific and engineering communities. The Council is administered jointly by both Academies and the Institute of Medicine. Dr. Bruce Alberts and Dr. Robert M. White are chairman and vice chairman, respectively, of the National Research Council.

This material is based on work supported by the National Institutes of Health.

Library of Congress Catalog Card Number 94-66477
International Standard Book Number 0-309-05086-3

Additional copies of this report are available from:

National Academy Press
2101 Constitution Avenue, NW
Box 285
Washington, DC 20055
800/624-6242
202/334-3313 (in the Washington Metropolitan Area)

B-452

Copyright 1994 by the National Academy of Sciences. All rights reserved.

Printed in the United States of America.

NATIONAL ACADEMY OF SCIENCES

2101 CONSTITUTION AVENUE, NW   WASHINGTON, D.C. 20418

OFFICE OF THE PRESIDENT

June 1, 1994

The Honorable Donna E. Shalala
Secretary of the Department of
    Health and Human Services
200 Independence Avenue, SW
Room 615-F
Washington, D.C. 20201

Dear Secretary Shalala:

It is a pleasure to present to the Department of Health and Human Services a copy of the 1994 report of the Committee on National Needs for Biomedical and Behavioral Research Personnel. This is the tenth in a series of reports undertaken by the National Research Council pursuant to Title I of the National Research Act of 1974 (P.L. 93-348 as amended). The work has been supported under Contract NO1-OD-2-2116/C with the National Institutes of Health.

The Act states that the purposes of the continuing study are to: "(1) establish (A) the Nation's overall need for biomedical and behavioral research personnel, (B) the subject areas in which such personnel are needed and the number of such personnel needed in each such area, and (C) the kinds and extent of training which should be provided such personnel; (2) assess (A) current training programs available for the training of biomedical and behavioral research personnel that are conducted under this Act at or through the institutes...and (B) other current training programs available for the training of such personnel; (3) identify the kinds of research positions available to and held by individuals completing such programs; (4) determine, to the extent feasible, whether the programs referred to in clause (B) of paragraph (2) would be adequate to meet the need established under paragraph (1) if the programs referred to in clause (A) of paragraph (2) were terminated; and (5) to determine what modifications in the programs referred to in paragraph (2) are required to meet the needs established under paragraph (1)."

Previous NRC reports have provided guidance to the NIH and the Alcohol, Drug Abuse and Mental Health Administration and to the U.S. Congress about the appropriate size and composition of the NRSA program given national needs for these highly skilled scientists. In addition to the core activities stipulated by the National Research Act, the agency directed the NRC to review the mathematical projection models of supply and demand used by previous NRC study committees and to establish their adequacy in addressing "national needs" issues in the 1990s.

The Honorable Donna E. Shalala
Page Two

    This report includes a new approach to modeling supply which should be regarded as exploratory. As a result, the Committee's recommendations for award levels are based more heavily on expert judgment than has been the case in the past. The Committee's Panel on Estimation Procedures, I should add, will prepare a report for release later this year which addresses the more general matter of mathematical approaches to the estimation of "need".

    Through a combination of a variety of information gathering activities and Committee deliberations, the Committee has concluded that the nation's need for these scientists remains strong and that the NRSA program, while small compared to the many other sources of doctoral and postdoctoral support, is enormously powerful in terms of its ability to change research emphases and to attract the highest quality individuals to research careers in the basic biomedical, behavioral and clinical sciences. The Committee has described in this report the next steps that are needed to assure that the NRSA program fulfills its intended role in fostering and maintaining a strong human resource base for health research.

    We hope the present report will be helpful and would be pleased to discuss it with you and your staff.

Sincerely,

Bruce M. Alberts
President

Enclosure

NATIONAL RESEARCH COUNCIL
Office of Scientific and Engineering Personnel
Studies and Surveys Unit

COMMITTEE ON NATIONAL NEEDS FOR
BIOMEDICAL AND BEHAVIORAL RESEARCH PERSONNEL

Ira J. Hirsh, *Co-chair*
Washington University (Retired)
Central Institute for the Deaf

John D. Stobo, *Co-chair*
Department of Medicine
The Johns Hopkins University

Helen M. Berman
Department of Chemistry
Rutgers University

Francis J. Bullock
Arthur D. Little, Inc.

Edwin C. Cadman
Department of Medicine
Yale University School of Medicine

Nancy E. Cantor
Department of Psychology
Princeton University

Eli Ginzberg
Eisenhower Center for the Conservation of
  Human Resources
Columbia University

Robert Hill*
Department of Biochemistry
Duke University Medical Center

R. Duncan Luce
Institute for Mathematical and Behavioral Sciences
University of California at Irvine

Ruth McCorkle
School of Nursing
University of Pennsylvania

Raymond Nickerson
Bolt Beranek & Newman (Retired)

Mary J. Osborn
Department of Microbiology & Biology
University of Connecticut Health Center

Cecil Payton
Department of Microbiology & Biology
Morgan State University

Richard Ranney
School of Dentistry
University of Maryland at Baltimore

Michael Rothschild
Division of Social Sciences
University of California at San Diego

Donald Steinwachs
Department of Health Policy and Management
Johns Hopkins University

Richard Thompson
Program for Neural, Informational, and Behavioral
  Sciences
University of Southern California

---

* Resigned March 1994.

COMMITTEE ON NATIONAL NEEDS FOR
BIOMEDICAL AND BEHAVIORAL RESEARCH PERSONNEL

PANEL ON ESTIMATION PROCEDURES

Michael Rothschild, *Chair*
Division of Social Sciences
University of California at San Diego

Eugene Hammel
Department of Demography
University of California at Berkeley

Robert Mare
Center for Demography and Ecology
University of Wisconsin

Alan Krueger
Woodrow Wilson School of Public and
  International Affairs
Princeton University

Aage Sørensen
Department of Sociology
Harvard University

NRC PROJECT STAFF

Alan Fechter, Executive Director
Office of Scientific and Engineering Personnel

Pamela Ebert Flattau
Director, Studies and Surveys Unit

Jeffrey E. Kallan
Staff Officer

Elaine Lawson
Research Associate

Anne L. Gallagher
Administrative Assistant

NATIONAL RESEARCH COUNCIL
OFFICE OF SCIENTIFIC AND ENGINEERING PERSONNEL

ADVISORY COMMITTEE

Linda S. Wilson, *Chair*
President
Radcliffe College

Ernest Jaworski, *Vice Chair*
Monsanto Company (Retired)

Betsy Ancker-Johnson
Chairman
World Environment Center

David Breneman
Graduate School of Education
Harvard University

David L. Goodstein
Vice Provost
Professor of Physics and Applied Physics
California Institute of Technology

Lester A. Hoel
Hamilton Professor of Civil Engineering
Duke University

Juanita M. Kreps
Department of Economics
Duke University

Donald Langenberg
Chancellor
University of Maryland System

Judith S. Liebman
Department of Chemical and Industrial Engineering
University of Illinois at Urbana-Champaign

Barry Munitz
Chancellor
The California State University

Kenneth Olden
Director, National Institute of Environmental
  Health Sciences
National Institutes of Health

Ewart A.C. Thomas
Department of Psychology
Stanford University

Annette B. Weiner
Dean, Graduate School of Arts and Sciences
New York University

William H. Miller *(Ex-officio)*
Department of Chemistry
University of California

PREFACE

In 1994 we mark the twentieth anniversary of the National Research Act of 1974 (P.L. 93-348), which established the National Research Service Awards (NRSA) program. Intended from the outset to augment federal programs of research support, the NRSA program was designed to increase the capability of the National Institutes of Health (NIH) and the Alcohol, Drug Abuse and Mental Health Administration (ADAMHA) to maintain a "superior program of research into the physical and mental disease and impairments of man...." Through a combination of training grants to institutions and the direct support of qualified individuals as research fellows, the NRSA program remains a significant force in the health research effort.

We cannot emphasize too strongly the significant impact the NRSA program has had on the federal system of predoctoral and postdoctoral training at U.S. universities. The 1974 legislation repealed existing research training and fellowship authorities of NIH and ADAMHA—one of which dated to the National Cancer Act of 1937— and consolidated research training under a single, new authority. In other words, the National Research Act of 1974 established a coherent system of support for recruiting individuals into health research and launching them into productive careers. Coupled with a variety of mechanisms to support training and education at all stages of the scientific career—from high school through midcareer—NIH provides the largest research training effort in the federal government, the centerpiece of which is the National Research Service Award.

In its 20 years of operation, the NRSA program has made it possible for many thousands of talented individuals in the basic biomedical, behavioral, and clinical sciences to sharpen their research skills and to apply those skills to topics of special concern to the nation, such as: aging, hypertension, the genetic basis of disease, acquired immune deficiency syndrome (AIDS), cancer, environmental toxicology, nutrition and health, and substance abuse. Surprisingly, few systematic studies are available of the career outcomes of NRSA trainees and fellows. Studies that are available, however, consistently note a distinctive role for NRSA trainees and fellows in the national health research effort. Nonetheless, many questions remain about the career outcomes of NRSA trainees and fellows, and it is our hope and that of our colleagues on the committee that NIH place high priority in the coming years on the careful analysis of career outcomes of NRSA graduates, determining to the extent possible the contributions of the NRSA program to health research relative to other forms of federal and private support for research training.

The continuing need for highly trained specialists to conduct research to meet the health needs of the country is as great today as it was 20 years ago. However, because of changes in patterns of research funding and the structure of the marketplace, the nature of this need has changed somewhat in recent years. Today there is a greater demand than in the past for talented health scientists to provide leadership in industrial research settings, in federal government laboratories, and in hospitals and clinics. The NRSA program continues to play a critical role in the preparation of many of those scientists.

It was within the context of these changing research opportunities that the National Research Council (NRC) agreed in 1992 to undertake this study of the NRSA program. In December 1992, the NRC appointed the Committee on National Needs for Biomedical and Behavioral Research Personnel, which we have been privileged to chair. It was our committee's task to establish the nation's overall need for biomedical and behavioral research personnel, the subject areas in which such personnel are needed, and the number of such personnel needed in each area for 1994 and beyond.

PREFACE

The committee was assisted by the Panel on Estimation Procedures, ably chaired by Michael Rothschild and staffed by Jeffrey Kallan. Through their careful collection and review of available statistics and various mathematical models, the panel provided us with a refreshing, alternative look at degree production and employment patterns in the many fields addressed by this study. The work of the panel continues beyond this report. We look forward with interest to the outcome of their deliberations, which should provide us with further insights into possible new approaches for assessing the nation's need for biomedical and behavioral scientists.

The work of the panel was augmented by information gathered by the committee through a public hearing and a series of commissioned papers. We are indebted to the many experts who offered valuable suggestions about possible new directions for the NRSA program.

In addition to these contributors, a number of people ensured a successful outcome of our efforts. Walter Schaffer, Research Training and Research Resources Officer at NIH, skillfully offered important information about the history and status of the NRSA program in his capacity as project officer. Dr. Schaffer arranged for numerous briefings by his colleagues at NIH. We are especially grateful to Ruth Kirschstein, Deputy Director of the NIH, who met with the committee in its early stages of discussion and offered helpful comments about areas of special concern to the NIH. We would also like to thank members of the NIH project oversight team for the information they provided, including Drs. David Chananie, Suzanne Feetham, Leonard Lash, James Lipton, John Norvell, James F. O'Donnell, and Carl Roth and Ms. Valerie Pickett.

The committee would also like to thank Carola Eisenberg who served as liaison from the NRC/OSEP Committee on Women in Science and Engineering and Ernest Jaworski who served as liaison from the Advisory Committee for the Office of Scientific and Engineering Personnel (OSEP). Both contributed in important ways to the deliberations of the committee. Kenneth Shine, President of the Institute of Medicine, also offered suggestions for analyses in the early stages in our work.

The committee would especially like to acknowledge the efforts of Alan Fechter, OSEP Executive Director. Together with Jeffrey Kallan, technical consultants Farrell Bloch and Peter Tiemeyer, and OSEP's data processing staff, most notably Marinus van der Have, Mr. Fechter effectively organized the labor force information contained in the basic biomedical and behavioral sciences chapters. The committee would also like to thank the staff of the Survey of Earned Doctorates and the Survey of Doctorate Recipients for their technical assistance, especially Delores Thurgood, Daniel Pasquini, and Prudence Brown. Pamela Ebert Flattau, Director of OSEP's Studies and Surveys Unit, guided the completion of this report and played a significant role in overseeing the coordination of the committee's overall study plan. Elaine Lawson contributed at key points in the study by gathering and summarizing a wide variety of material addressing national needs in the clinical sciences. Anne Gallagher, the committee's administrative assistant, worked tirelessly to coordinate the production of this volume. To these people, we express our gratitude for their efforts.

IRA J. HIRSH
JOHN D. STOBO, Co-Chairs
Committee on National Needs for
Biomedical and Behavioral Research Personnel

CONTENTS

EXECUTIVE SUMMARY   1
   Estimating National Needs For Research Scientists, 1
   Recommendations, 2
   Enhancing the Effectiveness of the NRSA Program, 8

1 INTRODUCTION   11
   Origins of the Study, 11
   Charge to the Committee, 12
   The Committee's Study Plan, 12
   Organization of the Report, 14

2 APPROACHES TO THE ESTIMATION OF NATIONAL NEED   17
   Health Research as a National Priority, 17
   Advances in Research, 19
   Marketplace Requirements, 20

3 BASIC BIOMEDICAL SCIENCES PERSONNEL   23
   Advances in Research, 24
   Assessment of the Current Market for
     Basic Biomedical Scientists, 25
   Outlook for Basic Biomedical Scientists, 30
   Ensuring the Diversity of Human Resources, 32
   The NRSA Program in the Basic Biomedical Sciences, 32
   Recommendations, 33

4 BEHAVIORAL SCIENCES PERSONNEL   37
   Advances in Research in the Behavioral Sciences, 38
   Assessment of the Current Market for Behavioral Scientists, 40
   Outlook for Behavioral Scientists, 45
   Ensuring the Diversity of Human Resources, 49
   The NRSA Program in the Behavioral Sciences, 49
   Recommendations, 50

5 PHYSICIAN-SCIENTISTS 55
Advances in Clinical Science, 56
Assessment of the Current Market for Clinical Scientists, 56
Outlook for Clinical Scientists, 57
Ensuring the Diversity of Human Resources, 59
The NRSA Program in the Clinical Sciences, 60
Recommendations, 62

6 ORAL HEALTH RESEARCH PERSONNEL 65
Advances in Oral Health Research, 65
Assessment of the Current Market for Oral Health Research Personnel, 65
Outlook for Oral Health Research Scientists, 66
Ensuring Diversity of Human Resources, 67
National Research Service Award Program for Oral Health Research, 67
Recommendations, 69

7 NURSING RESEARCH PERSONNEL 73
Advances in Nursing Research, 73
Assessment of the Current Market for Nursing Research Personnel, 75
Outlook for Nursing Research Personnel, 75
Ensuring Diversity of Human Resources, 76
The NRSA Program in Nursing Research, 76
Recommendations, 77

8 HEALTH SERVICES RESEARCH PERSONNEL 81
Advances in Health Services Research, 81
Assessment of the Current Market for Health Services Research Personnel, 83
Outlook for Health Services Research Personnel, 84
Ensuring Diversity of Human Resources, 86
The NRSA Program in Health Services Research, 86
Recommendations, 87

9 RECOMMENDATIONS AND REMAINING CONSIDERATIONS 89
Stipend Issues, 89
Enhancing the Effectiveness of the NRSA Program, 90

APPENDIXES
A Historical Overview 95
B Classification of Fields 103
C Public Hearing on National Needs for Biomedical and Behavioral Research Personnel 105
D Report Contributors 113
E Sources of Information for the National Research Service Award 115
F Data Tables 117
G Multistate Life Table Methodology and Projections 149
H Procedures Used to Estimate Awards, Stipends, and Costs 153
I Biographical Sketches 161

EXECUTIVE SUMMARY

The subject of this report is the nation's future need for biomedical and behavioral research scientists and the role the National Research Service Awards (NRSA) program can play in meeting those needs. The report has been prepared under the auspices of P.L. 93-348 (as amended), the National Research Act of 1974, which specifies that the Secretary of the Department of Health and Human Services shall request that the National Academy of Sciences conduct a continuing study of the nation's need for biomedical and behavioral scientists, the subject areas in which they are needed, and the kinds and extent of training which should be provided them. This is the tenth in the continuing series of reports to the National Institutes of Health (NIH) and the U.S. Congress on this subject.

The committee's deliberations and recommendations centered on three major activities. First, a Panel on Estimation Procedures was established to provide numerical analysis of educational and employment prospects and national needs for biomedical and behavioral research personnel. The panel concluded that existing mathematical models are inadequate to predict need and their recommendations for alternative strategies are presented in Chapter 2. Second, the committee convened a public hearing to gather views from a broad constituency associated with NRSA programs. Both written and oral testimony were considered by the committee and our conclusions are reflected in the recommendations that follow.[1] Third, the committee met as a group on four separate occasions and, as a panel of experts, discussed and debated information and recommendations. The consensus of these meetings forms the basis for the recommendations included in this report.

Through all the deliberations, one theme was constant: although the NRSA program may be relatively small as regards total numbers of trainees (less than 15 percent of the total number of graduate students training in the biomedical and behavioral sciences are supported by NRSA funds in any year), it is enormously powerful in its ability to change research emphases and to attract the highest quality individuals to research careers. It is viewed as a prestigious, highly competitive program. It is clear that initiatives, if introduced through the NRSA program, could have a powerful impact on intended new research directions or constituencies. In other words, the force of the NRSA program in determining research initiatives and directions in programs is enormous and clearly out of proportion to the relatively small place it occupies in the total research training portfolio.

ESTIMATING NATIONAL NEEDS FOR RESEARCH SCIENTISTS

The committee recognized early in its work that defining the nation's needs for biomedical and behavioral scientists is difficult and imprecise. However, the committee also recognized there are certain forces and opportunities that will have an impact on the needs for research personnel and, therefore, cannot be ignored. These include health care reform, major advances in fundamental research, and the demands of the marketplace. Recommendations for establishing the size and scope of the NRSA program represent the committee's best judgment concerning the continuation of a strong, health-related, scientific work force based on a consideration of these forces and opportunities.

Health Research as a National Priority

The expansion of national support for health research that occurred over the past few decades may be expected to continue although at a slower rate. Reapportionment of research and development funding between academia and industry has had and will continue to have an important effect on employment prospects for biomedical and behavioral scientists, although the effect will vary by field, as we discuss elsewhere in this report.

We expect continued national support for basic biomedical and behavioral research. However, research supported by federal funds may be expected to be closely relevant to social and economic needs and more readily adaptable for use in the private industrial sector. Furthermore, as society shifts away from product-intensive industries toward a service-oriented industrial base, the role of research in improving the quality and reducing the costs of health services will be closely scrutinized.

Unknown Effects of Health Care Reform

As forces converge to shape and change the delivery of health care in the United States, we expect a dramatic effect on research training. On the one hand, the potential flattening of reimbursement for clinical care will curtail a revenue stream (clinical income) that, in many medical schools and academic health centers, subsidizes the support for research training. A reduction in this revenue stream may shrink research training opportunities in academic health centers. On the other hand, increased emphasis on the maintenance of health, the outcome and quality of care, and the assessment of the impact of technology provide unique opportunities for prevention and health services research. Other changes in health care reform stress the need for increased activity in areas related to behavioral, nursing, and health services research.

Advances in Research

It is difficult, of course, to gauge the effect of the advancement of science on personnel needs. On the basis of our familiarity with the goals of the NRSA program, the range of expertise among our committee members, and the contributions of many individuals participating in our deliberations throughout the year, we have identified some fruitful areas of inquiry that also have the potential of engaging young scientists in careers that are both productive and rewarding.

Major advances in basic biomedical and behavioral research are ripe for application to address the causes, diagnosis, and treatment of human disease. These major research findings also provide the opportunity to bring new technologies to the marketplace.

Demand of the Marketplace

Employment conditions for biomedical and behavioral scientists were relatively robust throughout the 1980s. Dramatic changes have occurred, however, with regard to sector of employment with a greater fraction of Ph.D.s employed in industry and other nonacademic jobs than in earlier years.

The nation's need for research scientists has also been affected by demographic changes: the number of individuals from racial and ethnic minority groups is increasing but not as fast as might be expected given federal efforts to encourage the participation of minorities in this area. The work force of the future will consist of an increasing proportion of women and minorities; it is important that these changes are reflected in the biomedical and behavioral science work force.

RECOMMENDATIONS

Considering the major forces that have an impact on national needs for research and the unfolding of the research career, the committee recommends the following changes in the NRSA program to meet those needs.

Stipends

Raise the real value of stipends to more competitive levels by fiscal 1996: approximately $12,000 per year for predoctoral awardees and approximately $25,000 for postdoctoral awardees with less than 2 years of research experience. Maintain the real value of these stipends (i.e., the nominal value adjusted for inflation) through annual increases of 3 percent per year (the assumed annual rate of inflation).

It is disturbing to note that stipend levels for predoctoral trainees in the NRSA program have remained unchanged since 1991 at $8,800 taxable salary per year. The existing structure of a $700 monthly stipend is simply not sufficient. Many state university stipends start at $11,000 and the National Science Foundation currently pays $14,000. Hence, we recommend an increase in the inflation-adjusted value of predoctoral stipends to $12,000 by fiscal 1996 to provide an incentive for graduate students not only to seek but also to complete training at the doctoral level.

Postdoctoral NRSA awardees do not fare much better, earning approximately $18,600 in their first year of training and $19,700 in their second. It becomes very difficult at this important period of training to entice a clinician or Ph.D., already burdened with debt, into research training. Thus, the committee recommends that the NRSA stipends at the first-year postdoctoral level be increased to $25,000 in inflation-adjusted dollars by fiscal 1996. This expansion in stipend support should be achieved through the addition of funds to the current NRSA training budget (Appendix H).

EXECUTIVE SUMMARY

Numerical Recommendations

Between fiscal 1991 and 1993, the total number of NRSA awards grew from just over 14,000 to over 15,000 (Summary Table 1). The basic biomedical sciences were estimated to have the largest fraction of support in fiscal 1993 at about 9,633 awards, followed by 3,000 awards in the clinical sciences [excluding 822 awards for combined M.D.-Ph.D. training through the Medical Scientist Training Program (MSTP)]. To meet the nation's future needs for biomedical and behavioral scientists, we believe the overall NRSA program should expand from 15,112 slots in fiscal 1993 to 16,260 slots in fiscal 1996, with that growth occurring mainly through modest expansion of NRSA support for research training through the MSTP program, in the behavioral sciences, oral health research, nursing research, and health services research (Summary Table 2). With full implementation of the recommendations that follow the number of NRSA awards would have expanded by about 10 percent between 1993 and 1996 with support for basic biomedical sciences representing 59 percent of the total in 1996, behavioral sciences about 9 percent of the total, the MSTP program 6 percent, other clinical sciences (including oral health research) 20 percent, nursing research 3 percent, and health services research 2 percent.

Recommended changes in the size and scope of the NRSA program within each broad field are summarized below.

Basic Biomedical Sciences

Maintain the annual number of predoctoral awards in the basic biomedical sciences at 1993 levels, or approximately 5,175 awards, and the number of postdoctoral awards at 3,835.

On the basis of input from a wide variety of sources about current and anticipated market conditions and in consideration of pressing national research needs, the committee endorses the continuation of federal support through predoctoral awards in the basic biomedical sciences (see Chapter 3). The committee is concerned, however, that the current low levels of stipend support will not attract the most talented students to careers in research. To underscore the depth of our concerns, we recommend that predoctoral awards in the basic biomedical sciences be maintained at fiscal 1993 levels until further assessment of funding priorities and national needs can be made. The committee recognizes that these recommendations are made in an era of fiscal restraint. Should additional funds become available for predoctoral research training in the basic biomedical sciences, NIH might wish to consider expanding NRSA support in this area.

Postdoctoral research training is also an important component in the preparation of productive investigators in the basic biomedical sciences. Postdoctoral training increases the technical skills of the doctoral-level scientist and ensures the success of their independent research careers. Here, too, the committee is concerned that persistent low-level stipends may discourage qualified applicants from seeking postdoctoral training through the NRSA support. Thus, to permit NIH to introduce further and more realistic changes in stipend levels at the postdoctoral level, the committee recommends that the number of postdoctoral awards be maintained at fiscal 1993 levels. Again, however, should additional program funds become available for postdoctoral research training in the basic biomedical sciences, NIH might wish to increase the number of these awards.

Behavioral Sciences

Increase the annual number of NRSA awards for research training in the behavioral sciences from 1,069 to 1,450 between 1993 and 1996.

On the basis of continuing gains being made by behavioral scientists in areas of national interest and on anticipated demand for behavioral research relative to health goals, the committee urges the continued expansion of federal support through predoctoral awards in the behavioral sciences (see Chapter 4). Predoctoral awards permit the preparation of investigators familiar with the broad range of research techniques and theories that characterize doctoral preparation in the behavioral sciences. As is the case in other areas, the committee is concerned that current low stipend levels for NRSA awardees do not attract the most able scientists to research careers in health-related fields. Thus, the committee has tempered its call for expansion in total support from 672 predoctoral awards in fiscal 1993 to 900 by fiscal 1996 in recognition of the competing need to increase stipend support.

Postdoctoral research training through the NRSA provides the nation with an unusual mechanism for attracting the most skilled scientists to address areas of national need. Because of differences in the evolution of research careers, postdoctoral research training has played a greater role in some behavioral science fields than others. Nonetheless, postdoctoral study increases the technical skills of the investigator and strengthens the pool of talent available to the nation for research. Thus, the committee recommends that the number of postdoctoral trainees and fellows supported annually in the behavioral sciences increase from approximately 349 awardees in fiscal 1993 to 500 in fiscal 1996.

Clinical Sciences

Increase the number of MSTP awards from 822 in 1993 to 1,020 by 1996 and the number of postdoctoral

SUMMARY TABLE 1 Aggregated numbers of NRSA supported trainees and fellows for FY 1991, FY 1992, and FY 1993.[a]

Fiscal Year	Type of Program	TOTAL ALL FIELDS	Basic Biomedical Sciences	Behavioral Sciences	Clinical Sciences	Medical Scientist[b] Training	Oral Health[c] Research	Nursing[d] Research	Health Services[e] Research
1991 TOTAL	Total	14,085	9,021	902	2,894	783	218	255	12
	Predoctoral	6,948	4,593	519	755	783	78	220	0
	Postdoctoral	6,525	3,861	338	2,139	0	140	35	12
	MARC Undergraduate	612	567	45	0	0	0	0	0
Trainees	Total	11,850	7,199	775	2,814	783	186	93	0
	Predoctoral	6,449	4,313	472	736	783	78	67	0
	Postdoctoral	4,789	2,319	258	2,078	0	108	26	0
	MARC Undergraduate	612	567	45	0	0	0	0	0
Fellows	Total	2,235	1,822	127	80	0	32	162	12
	Predoctoral	499	280	47	19	0	0	153	0
	Postdoctoral	1,736	1,542	80	61	0	32	9	12
1992 TOTAL	Total	14,607	9,317	908	2,970	806	213	257	94
	Predoctoral	7,265	4,777	534	819	806	77	217	35
	Postdoctoral	6,661	3,910	323	2,151	0	136	40	59
	MARC Undergraduate	681	630	51	0	0	0	0	0
Trainees	Total	12,365	7,477	790	2,887	806	178	103	82
	Predoctoral	6,761	4,487	481	800	806	77	75	35
	Postdoctoral	4,923[f]	2,360	258	2,087	0	101	28	47
	MARC Undergraduate	681	630	51	0	0	0	0	0
Fellows	Total	2,242	1,840	118	83	0	35	154	12
	Predoctoral	504	290	53	19	0	0	142	0
	Postdoctoral	1,738	1,550	65	64	0	35	12	12
1993 TOTAL	Total	15,112	9,633	1,069	2,974	822	224	236	96
	Predoctoral	7,835	5,171	672	855	822	97	188	30
	Postdoctoral	6,603	3,836	349	2,119	0	127	48	66
	MARC Undergraduate	674	626	48	0	0	0	0	0
Trainees	Total	12,819	7,740	930	2,877	822	201	112	79
	Predoctoral	7,265	4,811	604	826	822	96	76	30
	Postdoctoral	4,880[g]	2,303	278	2,051	0	105	36	49
	MARC Undergraduate	674	626	48	0	0	0	0	0
Fellows	Total	2,293	1,893	139	97	0	23	124	17
	Predoctoral	570	360	68	29	0	1	112	0
	Postdoctoral	1,723	1,533	71	68	0	22	12	17

NOTE: Data from IMPAC data system was prepared by the information and Statistics Branch, Division of Research Grants, and the National Institues of Health, RTSPO/OEP/OER. About 230 positions in FY 1991 and 576 positions in FY 1992 were not coded for disciplines in IMPAC datafiles. These positions were assigned to cluster disciplines using departmental affiliations and grant titles.

a Units are full time training positions (FTTPs). Short term training positions are included, but the number of appointments have been divided by four to convert to FTTPs.

b Positions designated as Medical Scientist Training Program (MSTP) for purposes of this table are all considered to be in biomedical disciplines.

c Positions supported by the National Institute of Dental Research (NIDR).

d Positions supported by the National Institute of Nursing Research (NINR).

e Positions supported by the Agency for Health Care Policy and Research (AHCPR).

f Includes 42 postdoctoral traineeships in 1992 for training in Primary Care Research supported through the Health Resources and Services Agency (HRSA).

g Includes 58 postdoctoral traineeships in 1993 for training in Primary Care Research supported through the Health Resources and Services Agency (HRSA).

fellows in the clinical sciences from 68 in 1993 to 160 in 1996. To achieve this expansion, we recommend that the annual number of postdoctoral trainees in the clinical sciences be decreased slightly from 2,051 to 1,805 between 1993 and 1996.

Studies have consistently shown that a substantial fraction of graduates from the MSTP program remain productively engaged in research, often with greater success in securing research support than those M.D.s who pursue post-M.D. research training not leading to a doctorate (see Chapter 5). Current support for MSTP training provides for about 820 awards. Given the success of this program in contributing workers to the national research effort, we believe this program should be expanded to provide 1,020 awards by fiscal 1996.

Furthermore, because of the urgent need for clinical scientists familiar with patient-based research techniques we urge the NIH to increase the number of postdoctoral NRSA fellowship awards from 68 in fiscal 1993 to 160 by fiscal 1996 to permit the preparation of patient-based investigators.

To permit the expansion of the pool of MSTP trainees and postdoctoral fellows, we believe modest reductions should be made in the number of postdoctoral awards made through institutional training grants in the clinical sciences. NIH reports that 2,051 awardees were supported in fiscal 1993 through this mechanism. We believe a gradual decrease in the number of awards to 1,805 should occur by fiscal 1996.

Remaining Fields

Increase the number of awards in oral health research to 430 by 1996, in nursing research to 500 by 1996, and in health services research to 360 by 1996, to allow for their efficient absorption by the system after which the number of awards is stabilized at the new levels.

Oral Health Research. There is an acute need for clinical dental researchers and oral health research workers in general. The National Research Council's 1985 report called for 320-400 new clinical dental research trainees annually, but the National Institute of Dental Research (NIDR) has been unable to carry out this recommendation because funds have not been available. A significant increment in training would substantially alleviate the shortage of oral health research personnel. There is need and rationale for a tripling or quadrupling the training of oral health researchers. Realistically, however, the need is better met incrementally rather than abruptly. The committee recommends, therefore, that the total number of training positions available for preparation in oral health research increase from approximately 224 positions in fiscal 1993 to 430 positions in fiscal 1996 and remain steady thereafter (see Chapter 6).

The MSTP program offers an integrated program of medical and graduate research training leading to the combined M.D. and Ph.D. degrees. The success of that program, coupled with the demonstrated success of such joint degrees as the D.D.S./Ph.D. and D.M.D./Ph.D., suggests that oral health research would benefit from the development of a Dental Scientist Training Program (DSTP) that is analogous to the MSTP under the auspices of the NRSA legislation. The committee recommends, therefore, that one-quarter to one-half of the new positions available for training in oral health research in fiscal 1994 and beyond be used by NIDR to establish a DSTP program under the NRSA act.

Nursing. With the proposed changes in health care reform, continued development of a strong scientific base in nursing research for practice is essential to prepare advanced practice specialists to care for the rapidly changing needs of high-risk and underserved patient populations (see Chapter 7). The committee recommends that the number of positions available for preparation in nursing research increase from approximately 236 awards in fiscal 1993 to 500 in fiscal 1996. These positions should be phased in on a yearly basis as properly qualified candidates and training sites present themselves.

Because nursing research is a developing field of science, there is a critical need to have an increased number of highly trained nurse researchers at the cutting edge for nursing practice and health care. Support for research training must be expanded at the predoctoral level to allow further expansion at the postdoctoral level. As the number of NRSA positions increase by the year 1996, there should be a progressive shift toward an eventual balance between the proportion of funding for predoctoral and postdoctoral support.

Health Services Research. Health services research is a relatively young field that uses interdisciplinary approaches to examine the impact of organization, finance, and use of technology on the utilization, cost, and quality of care (see Chapter 8). This field of research will need to grow substantially to meet the ever-expanding demands for information by policymakers, administrators, providers, and consumers. The questions raised regarding what impact different proposals for health care reform will have on access, cost, and quality of care are largely questions that will be addressed by this field of research. To meet these needs, the committee recommends that the number of NRSA positions allocated to the Agency for Health Care Policy and Research (AHCPR) increase from about 96 in fiscal 1993 to 360 in fiscal 1996.

Institutional training grants permit the development of

SUMMARY TABLE 2 Committee Recommendations for NRSA Trainees and Fellows for FY 1994 through 1999.[a]

Fiscal Year	Type of Program	TOTAL ALL FIELDS	Basic Biomedical Sciences	Behavioral Sciences	Clinical Sciences	Medical Scientist Training [b]	Oral Health Research [c]	Nursing Research [d]	Health Services Research [e]
1994 REC.	Total	15,415	9,640	1,195	2,975	890	260	340	115
	Predoctoral	8,175	5,175	745	895	890	125	290	55
	Postdoctoral	6,560	3,835	400	2,080	0	135	50	60
	MARC Undergraduate	680	630	50	0	0	0	0	0
Trainees	Total	12,975	7,745	1,040	2,875	890	200	130	95
	Predoctoral	7,490	4,815	670	875	890	100	95	45
	Postdoctoral	4,805	2,300	320	2,000	0	100	35	50
	MARC Undergraduate	680	630	50	0	0	0	0	0
Fellows	Total	2,440	1,895	155	100	0	60	210	20
	Predoctoral	685	360	75	20	0	25	195	10
	Postdoctoral	1,755	1,535	80	80	0	35	15	10
1995 REC.	Total	15,835	9,640	1,325	2,910	955	345	420	240
	Predoctoral	8,600	5,175	825	895	955	210	360	180
	Postdoctoral	6,555	3,835	450	2,015	0	135	60	60
	MARC Undergraduate	680	630	50	0	0	0	0	0
Trainees	Total	13,165	7,745	1,150	2,780	955	230	160	145
	Predoctoral	7,730	4,815	740	875	955	130	120	95
	Postdoctoral	4,755	2,300	360	1,905	0	100	40	50
	MARC Undergraduate	680	630	50	0	0	0	0	0
Fellows	Total	2,670	1,895	175	130	0	115	260	95
	Predoctoral	870	360	85	20	0	80	240	85
	Postdoctoral	1,800	1,535	90	110	0	35	20	10
1996 REC.	Total	16,260	9,640	1,450	2,860	1,020	430	500	360
	Predoctoral	9,010	5,175	900	895	1,020	290	430	300
	Postdoctoral	6,570	3,835	500	1,965	0	140	70	60
	MARC Undergraduate	680	630	50	0	0	0	0	0
Trainees	Total	13,355	7,745	1,260	2,680	1,020	265	195	190
	Predoctoral	7,965	4,815	810	875	1,020	160	145	140
	Postdoctoral	4,710	2,300	400	1,805	0	105	50	50
	MARC Undergraduate	680	630	50	0	0	0	0	0
Fellows	Total	2,905	1,895	190	180	0	165	305	170
	Predoctoral	1,045	360	90	20	0	130	285	160
	Postdoctoral	1,860	1,535	100	160	0	35	20	10

a Units are full time training positions (FTTPs). Short term training positions are included, but the number of appointments have been divided by four to convert to FTTPs.

b Positions designated as Medical Scientist Training Program (MSTP) for purpose of this table are all considered to be in biomedical disciplines.

EXECUTIVE SUMMARY

SUMMARY TABLE 2 (continued).

Fiscal Year	Type of Program	TOTAL ALL FIELDS	Basic Biomedical Sciences	Behavioral Sciences	Clinical Sciences	Medical Scientist[b] Training	Oral Health[c] Research	Nursing[d] Research	Health Services[e] Research
1997 REC.	Total	16,260	9,640	1,450	2,860	1,020	430	500	360
	Predoctoral	9,010	5,175	900	895	1,020	290	430	300
	Postdoctoral	6,570	3,835	500	1,965	0	140	70	60
	MARC Undergraduate	680	630	50	0	0	0	0	0
Trainees	Total	13,355	7,745	1,260	2,680	1,020	265	195	190
	Predoctoral	7,965	4,815	810	875	1,020	160	145	140
	Postdoctoral	4,710	2,300	400	1,805	0	105	50	50
	MARC Undergraduate	680	630	50	0	0	0	0	0
Fellows	Total	2,905	1,895	190	180	0	165	305	170
	Predoctoral	1,045	360	90	20	0	130	285	160
	Postdoctoral	1,860	1,535	100	160	0	35	20	10
1998 REC.	Total	16,260	9,640	1,450	2,860	1,020	430	500	360
	Predoctoral	9,010	5,175	900	895	1,020	290	430	300
	Postdoctoral	6,570	3,835	500	1,965	0	140	70	60
	MARC Undergraduate	680	630	50	0	0	0	0	0
Trainees	Total	13,355	7,745	1,260	2,680	1,020	265	195	190
	Predoctoral	7,965	4,815	810	875	1,020	160	145	140
	Postdoctoral	4,710	2,300	400	1,805	0	105	50	50
	MARC Undergraduate	680	630	50	0	0	0	0	0
Fellows	Total	2,905	1,895	190	180	0	165	305	170
	Predoctoral	1,045	360	90	20	0	130	285	160
	Postdoctoral	1,860	1,535	100	160	0	35	20	10
1999 REC.	Total	16,260	9,640	1,450	2,860	1,020	430	500	360
	Predoctoral	9,010	5,175	900	895	1,020	290	430	300
	Postdoctoral	6,570	3,835	500	1,965	0	140	70	60
	MARC Undergraduate	680	630	50	0	0	0	0	0
Trainees	Total	13,355	7,745	1,260	2,680	1,020	265	195	190
	Predoctoral	7,965	4,815	810	875	1,020	160	145	140
	Postdoctoral	4,710	2,300	400	1,805	0	105	50	50
	MARC Undergraduate	680	630	50	0	0	0	0	0
Fellows	Total	2,905	1,895	190	180	0	165	305	170
	Predoctoral	1,045	360	90	20	0	130	285	160
	Postdoctoral	1,860	1,535	100	160	0	35	20	10

c Positions supported by the National Institute of Dental Research (NIDR).
d Positions supported by the National Institute of Nursing Research (NINR).
e Positions supported by the Agency for Health Care Policy and Research (AHCPR).

innovative interdisciplinary research training programs, an essential feature of research in this area. However, given the anticipated growing demand for skilled specialists in health services research, the committee concludes that AHCPR should place emphasis on the award of individual fellowships in the next few years in order to encourage qualified individuals with some experience in the area of health care policy to pursue advanced training.

Minority Access to Research Careers

Hold Minority Access to Research Careers (MARC) awards constant at fiscal 1993 levels, or approximately 680 awards, pending the outcome of the present NIH evaluation study.

The NRSA program plays an important role in attracting minority group members to careers in the basic biomedical and behavioral sciences. This is achieved primarily through the MARC program (see Chapters 3 and 4). The core of this program is the Honors Undergraduate Program launched in fiscal 1977 to support college juniors and seniors (see, for example, Garrison et al., 1985). In fiscal 1993 about 674 slots were set aside by NIH for preparing of MARC undergraduates, most for training in the basic biomedical sciences. The committee is aware that NIH recently launched an 18-month study of the career outcomes of MARC program graduates. The committee endorses this effort and recommends a continuation of support for the MARC program at fiscal 1993 levels until the NIH assessment is complete.

It is well documented that certain minority groups (African Americans, Hispanics, Native Americans) are vastly underrepresented in the biomedical and behavioral sciences. To address this problem, the obvious solution should be to increase opportunities for these groups to receive training in biomedical and behavioral sciences. This can only be accomplished by providing research training opportunities through such programs as MARC. It will be important that the NIH act swiftly to review and modify the size and scope of the MARC program in light of its findings.

In recent years, the NIH has undertaken a number of important studies of minority research and training.[2] For example, the Office of Minority Health in 1993 reported on Phase I of an assessment of minority training programs (NIH, 1993). This report lays out a plan of action for reviewing NIH programs in this area and discusses the limitations of current NIH data collection procedures for tracking individuals who receive research and/or training support.[3] The range of programs to be addressed by this assessment is impressive, and includes such programs as: Minority Biomedical Research Support (MBRS), National Predoctoral Fellowship Awards for Minority Students, Minority Access to Research Careers (MARC), Research Centers in Minority Institutions (RCMI), and Research Supplements for underrepresented minority individuals.

The committee looks forward to having more detailed information in the coming years on the outcome of these various assessments to facilitate the development of a fuller review of minority research training needs and the role of the NRSA program in meeting those needs.

ENHANCING THE EFFECTIVENESS OF THE NRSA PROGRAM

Flexibility in Career Training at the Postdoctoral Level

Examine research training opportunities for women through the NRSA program and strengthen the role of postdoctoral support to assist women in establishing themselves in productive careers as research scientists.

In May 1993 we convened a public hearing to invite suggestions for increasing the effectiveness of the NRSA program. Most of those testifying on the role of the NRSA program in recruiting women said that the NRSA program must be more flexible in the areas of reentry training, family leave, and geographic location of training sites. Committee members have also been concerned, however, that there is a clear disparity between the number of women receiving NRSA training and the number of recipients of NIH research grants (see Chapter 9).

Women appear to be leaving science between the time they receive their doctorate and the time that they fully establish themselves in a research career track. The NRSA program can play a role in fostering the careers of these scientists. There is a need, then, to reshape NRSA awards at the postdoctoral level to encourage women to fully utilize their research talents. NRSA awards should allow retraining and career reentry to help women who have stopped out of research to update skills and move into emerging areas.

Monitoring Progress Toward NRSA Goals

Review NIH databases as management information systems and introduce changes in data collection, analysis, and dissemination to permit more effective tracking of NRSA award recipients. Emphasis should be given to the analysis of minority participation in research and training. New funds should be directed to the evaluation of NRSA program outcomes.

Perhaps one of the most significant findings of this committee is the general lack of information about the outcome of the NRSA program given almost two decades of support. Very little serious evaluation of the NRSA program has been undertaken through NIH support except for a few student outcomes studies undertaken by earlier NRC committees. Nowhere is the need for accurate information more evident than in our inability to track the participation of underrepresented minorities in the biomedical and behavioral research effort. We cannot underscore strongly enough the need for follow-up information to assess program outcomes. In part, this involves the organization of existing files at NIH to permit the analysis of program outcomes (see Chapter 9). In part, the analysis that is needed will require serious review of data collection and analytic capabilities at the NIH and the development of new strategies to assess program outcomes.

Improving the effectiveness of the NRSA program will require attention to issues not new to the research community. However, with the inevitable changes that will occur with health care reform and budget deficit reduction, NIH may find itself in a position of justifying its support for training programs. Well-designed career outcomes studies can provide the kind of feedback that is needed to assure that the NRSA program is both efficient and effective given constraints being placed on the federal funding effort. Future committees would benefit, furthermore, from more studies of the impact of NRSA support on the recipient institution's total pattern of training support.

NOTES

1. The public testimony from the hearing will be available in a separate report, through the National Academy Press, Washington, D.C.

2. In addition to the studies described here, the Director of the National Institutes of Health has also commissioned a study of NIH efforts to recruit and retain minority scientists among its intramural staff. A report on that topic is being prepared by the Director's NIH/EEO Executive Advisory Group. (Personal communication, 1994)

3. The NIH Minority Programs Evaluation Committee recommended changes in the NIH data system to permit more effective tracking of individuals, similar to the changes recommended by this committee (see Chapter 9).

REFERENCES

Garrison, H.H., P.W. Brown and R.W. Hill
    1985   *Minority Access to Research Careers: An Evaluation of the Honors Undergraduate Research Training Program.* Washington, D.C.: National Academy Press.

National Institutes of Health (NIH)
    1993   *Assessment of NIH Minority Research/Training Programs: Phase* I. Bethesda, MD: National Institutes of Health.

CHAPTER ONE

INTRODUCTION

Every year a fraction of freshmen at U.S. colleges and universities indicate their interest in becoming research scientists (Dey et al., 1991). Fluctuating between 1 and 3 percent of each entering class since 1979, these students represent a small but important component of the talent pool from which the future leaders of U.S. science and technology will emerge. The early expression of interest in a science career comes after many years of exposure to science and mathematics in primary and secondary schools. It will be followed by many more years of careful preparation culminating in the attainment of a Doctor of Philosophy (Ph.D.) degree or other research doctorate.

Given the lengthy process of converting aspirants into researchers, it should come as no surprise that many changes take place in the size and composition of that talent pool as these students, and others, move into and out of the research training and career track. It is the fluid nature of this path toward a scientific career that has led the federal government to initiate programs and policies aimed at fostering the development of the human resource base in science and technology. Among the numerous sources of graduate and postdoctoral support, the National Research Service Awards (NRSA) program is unique. Through a competitive program of individual and institutional support, the NRSA program promotes the development of a pool of skilled scientists in specific areas of "national need." It is a program of support that traces its roots to early efforts by the federal government to link the development of research areas with training, the fundamental assumption being that the quality of the research enterprise depends on the talents of individuals attracted to a career in research (Lenfant, 1989). (See Appendix A for a brief history of the NRSA program.) The subject of this report is the future direction of that program given anticipated changes in the delivery of health care, exciting developments in health research, and important changes in the composition of the biomedical and behavioral sciences work force.

ORIGINS OF THE STUDY

For nearly 20 years, the National Research Council (NRC) has played an active role in the ongoing review of training opportunities available to individuals seeking advanced preparation in the biomedical and behavioral sciences. During this time, NRC has issued nine reports describing the optimal structure of the NRSA program given national requirements for health-related research scientists and available training opportunities.

The involvement of the NRC in the assessment of national needs for biomedical and behavioral research scientists originates in the National Research Service Award Act of 1974 (P.L. 93-348), which abolished all previous training authorities of the National Institutes of Health (NIH) and consolidated training into a single training authority. The legislation stipulated that these awards should be restricted to subject areas for which there is a need for personnel (see Box 1-1).

The same legislation directed the Secretary of Health and Human Services (as the department is known today) to arrange for a continuing study of "national needs" and to request that the National Academy of Sciences conduct a study that would, in part, establish the subject areas in which such personnel are needed (see Box 1-2).

Since 1975, the NRC—the principal operating arm of the National Academy of Sciences, the Institute of Medicine, and the National Academy of Engineering—has submitted nine reports to the Secretary in response to this request for advice. This report, the tenth in the series, responds to the Congressional charge and provides additional information requested by NIH.

> **BOX 1-1 The National Research Service Award Act of 1974 (P.L. 93-348)**
>
> **FINDINGS AND DECLARATION OF PURPOSE**
>
> Sec. 102. (a) Congress finds and declares that—
> (1) the success and continued viability of the Federal biomedical and behavioral research effort depends on the availability of excellent scientists and a network of institutions of excellence capable of producing superior research personnel;
> (2) direct support of the training of scientists for careers in biomedical and behavioral research is an appropriate and necessary role for the Federal Government; and
> (3) graduate research assistance programs should be the key elements in the training programs of the institutes of the National Institutes of Health and the Alcohol, Drug Abuse, and Mental Health Administration.
> (b) It is the purpose of this title to increase the capability of the institutes of the National Institutes of Health and the Alcohol, Drug Abuse, and Mental Health Administration to carry out their responsibility of maintaining a superior national program of research into the physical and mental disease and impairments of man ...
>
> **NATIONAL RESEARCH SERVICE AWARDS**
>
> Sec. 472. (a) ... (3) Effective July 1, 1975, National Research Service Awards may be made for research or research training in only those subject areas for which, as determined under section 473, there is a need for personnel ...

CHARGE TO THE COMMITTEE

Previous NRC reports have provided guidance to NIH and the Alcohol, Drug Abuse and Mental Health Administration (ADAMHA) relative to the appropriate size and composition of the NRSA program. Recommendations have been made on the number of research training positions to be supported, and these recommendations have been used by NIH and ADAMHA staff during the budget formulation and by the U.S. Congress during the appropriations process (NIH, 1992). Once again the agency requested that NRC estimate the future demand for researchers, estimate the current supply, and, using estimates of the future demand and knowledge of the current balance between supply and demand, make recommendations on the appropriate size of the NRSA program.

In addition to these core activities, NIH also requested assistance in assessing the effectiveness of the NRSA program by gathering and analyzing information on such issues as the adequacy of current stipends to attract talented individuals into research careers in the biomedical and behavioral sciences. Furthermore, the agency directed NRC to review the mathematical projection models of supply and demand used by previous NRC study committees and to establish their adequacy in addressing "national needs" issues in the 1990s. In response to this request, the committee developed a five-part study plan.

THE COMMITTEE'S STUDY PLAN

Definition of the Study Population

The first step in undertaking this study was to develop a list of fields that were understood to define each of six broad areas of training. The definition of the study population is critical to the success of this analytic effort because the field taxonomy (see Appendix B) establishes the categories for data analysis. Thus, the basic biomedical sciences are understood to include biochemistry, molecular biology, and the like, whereas the behavioral sciences include psychology, sociology, anthropology, and speech and hearing sciences. Field matching was then made possible among data bases maintained by NIH, the National Science Foundation, NRC, the Association of American Medical Colleges, and American Association of Colleges of Nursing, among others.

Field matching in the clinical sciences proved more difficult. As we embarked on the study, representatives of the dental research community indicated to us that they considered their field code too restrictive, not accurately reflecting the diverse nature of their research base. The committee, through its staff representatives, conducted extensive consultative sessions with staff of the National Institute for Dental Research (NIDR), and from these sessions the concept of "oral health research" emerged. A workshop was organized to assess the market for oral health research personnel and the results were incorporated in the committee's

INTRODUCTION

deliberations (see Chapter 6 of this report). Thus, individuals familiar with the taxonomy used by earlier NRC study committees should be aware that a new field designation has been developed for the area of research training formerly designated "dental science" research training.

With regard to the clinical sciences more broadly defined, the committee has concentrated its attention this year on the continuing need to recruit physician-scientists into the research career path. This critical group of scientific workers includes individuals holding medical doctorates and those with combined medical and research doctorates (see Chapter 5).

Because of the special market for nurses who pursue advanced preparation in research, this report includes a separate assessment of research training needs in nursing research (see Chapter 7). Nursing research personnel are generally defined as individuals holding both a degree in nursing and a Ph.D. Because of the wide variety of doctoral specialties pursued by these investigators, we have not specified degree or employment specialties of these researchers. Rather, we consider nursing research personnel to include individuals holding both a degree in nursing and a research doctorate in a wide variety of areas.

Finally, we have devoted a chapter in this report (Chapter 8) to a consideration of health services research personnel. Certain of the institutes of health provide research training in areas related to the improvement of health care delivery, such as drug abuse prevention studies. Our report concentrates, however, on the newly emerging market for research scientists stimulated by the establishment of the Agency for Health Care Policy and Research (AHCPR) in 1989. This agency now serves as the organizational locus within the federal government for studies of such health care reform issues as the reduction of health care costs, the quality of care for the aged, and the overall health status of Americans, drawing investigators from a wide variety of disciplines. Unfortunately, detailed information about the population of health services research personnel in the

---

BOX 1-2 Studies Regarding Biomedical and Behavioral Research Personnel (P.L. 93-348)

Sec. 473. (2) The Secretary shall, in accordance with subsection (b), arrange for the conduct of a continuing study to—

   (1) establish (A) the Nation's overall need for biomedical and behavioral research personnel, (B) the subject areas in which such personnel are needed and the number of such personnel needed in each such area, and (C) the kinds and extent of training which should be provided such personnel;
   (2) assess (A) current training programs available for the training of biomedical and behavioral research personnel which are conducted under this Act at or through institutes under the National Institutes of Health and the Alcohol, Drug Abuse, and Mental Health Administration, and (B) other current training programs available for the training of such personnel;
   (3) identify the kinds of research positions available to and held by individuals completing such programs;
   (4) determine, to the extent feasible, whether the programs referred to in clause (B) or paragraph (2) would be adequate to meet the needs established under paragraph (1) if the programs referred to in clause (A) of paragraph (2) were terminated; and
   (5) determine what modifications in the programs referred to in paragraph (2) are required to meet the needs established under paragraph (1).
(b) (1) The Secretary shall request the National Academy of Sciences to conduct the study required by subsection (a) under an arrangement under which the actual expenses incurred by such Academy in conducting such study will be paid by the Secretary. If the National Academy of Sciences is willing to do so, the Secretary shall enter into such an arrangement with such Academy for the conduct of such study.
   (2) If the National Academy of Sciences is unwilling to conduct such study under such an arrangement, then the Secretary shall enter into a similar arrangement with other appropriate nonprofit private groups or associations under which such groups or associations will conduct such study and prepare and submit the reports thereon as provided in subsection (c).
   (c) A report on the results of such study shall be submitted by the Secretary to the Committee on Interstate and Foreign Commerce of the House of Representative and the Committee on Labor and Public Welfare of the Senate not later than March 31 of each year...

United States was last available in the 1970s when NRC conducted a number of specialized surveys of these scientists (see, for example, NRC, 1977). In the absence of systematic information about the study population, we have drawn on membership information from the Association of Health Services Research (AHSR) to develop a profile of selected—if not key—portions of that work force. As the demand for these specialists continues to grow, we hope that future study committees will be in a better position to specify the overall size of the health services research labor force and trends in training and employment. In the meantime, we have elected to base our training recommendations on the obvious immediate need to address current low levels of training in health services research through an expansion of the NRSA program administered by AHCPR to meet anticipated demand.

Panel on Estimation Procedures

To review mathematical and other models of supply and demand, the committee convened a panel of experts whose work continues beyond preparation of this report. The Panel on Estimation Procedures was formed to evaluate the adequacy of current models for estimating training needs in the broad fields comprising this study and to recommend new directions for enhancing models and for improving data collection needed to implement models in the future. A preliminary summary of their work may be found in the next chapter.

Call for a Public Hearing

Another important component of this study effort was a 1-day Public Hearing convened by the committee in Washington, D.C. (Appendix C). The hearing, convened on May 3, 1993, was designed to solicit the views of colleagues in the scientific and educational communities as part of the process of developing recommendations for the future direction of the NRSA program. At the hearing, we posed the following four questions to our speakers:

- What is the most significant challenge we face today in the United States in maintaining an adequate supply of qualified scientists to sustain and advance health research?
- What improvements might be made in the NRSA program to ensure a continuing supply of skilled investigators in the biomedical and behavioral sciences in the coming years?
- What steps might be taken to improve the effectiveness of the NRSA program in recruiting women and minorities into scientific careers?
- What features of the NRSA training grant might be strengthened to ensure the maintenance of high-quality research training environments?

The public hearing provided a forum for exploring the need for biomedical and behavioral research personnel in nontraditional settings, such as industry. It also provided a forum for interested spokesmen from the scientific and educational research communities to suggest modifications in the NRSA program. In preparation for the public hearing, the committee consulted with NIH and developed a comprehensive list of about 2,000 individuals and professional organizations to whom letters of invitation were sent. These letters requested brief abstracts of proposed presentations. Approximately 200 responses to the invitation were received, and from these approximately 35 speakers were selected. A summary of the public hearing will be available separately. Names of respondents are listed in Appendix C.

Commissioned Papers

To augment the expertise of the committee in a variety of areas, we commissioned a number of papers for use during our deliberations. In most cases, commissioned papers by experts within each of those broad fields addressed in this report augmented the quantitative analyses conducted under the guidance of the Panel on Estimation Procedures. Appendix D includes a list of the authors and other key contributors.

Airlie House Retreat

To review the information gathered by the committee through these various activities, the committee convened a three-day retreat in Airlie, Virginia, in September 1993. The purpose of the retreat was to review information gathered by the Panel on Estimation Procedures and other sources, to formulate our recommendations for refining the NRSA program, and to begin the task of organizing this report. Representatives from the Institute of Medicine, from the NRC Commission on Life Sciences, from the NRC Commission on Behavioral and Social Sciences and Education, and authors of the commissioned papers were present to assist the committee in their technical discussions.

ORGANIZATION OF THE REPORT

This report is organized into three parts. Chapter 2 summarizes our approach to the assessment of national need in this tenth report to the Secretary of Health and Human Services and to the U.S. Congress. This summary is followed by chapters addressing developments in the demand for research specialists in

- the basic biomedical sciences (Chapter 3),
- the behavioral sciences (Chapter 4),
- the clinical sciences (Chapter 5),

- oral health research (Chapter 6),
- nursing research (Chapter 7),
- health services research (Chapter 8).

Recommendations regarding the future direction of the NRSA program in meeting national needs are provided for each broad field (see Appendix E for a summary of the awards available through the NRSA program).

The report concludes with a consideration of overall training issues, including specification of steps that might be taken to make the NRSA program more effective in the coming years.

We are aware that there are certain limitations to this report. For example, we have not had the opportunity to address research training needs in many deserving areas, such as the clinical specialties of veterinary sciences or social work. We hope that subsequent analyses of personnel and training needs by the NRC will consider these and other topics not included in this volume. Furthermore, the committee was unable to collect new data about the biomedical and behavioral sciences work force given time and resource constraints. We believe, however, that this report serves the national interest by emphasizing specialty areas for which the demand for talent is great and for which information is generally available. We hope that NIH will consider collecting information in the near future needed to prepare for the next report, such as that described in Chapter 9.

REFERENCES

Dey, E. L., A. Astin and W. S. Korn
 1991 *The American Freshman: Twenty-Five Year Trends: 1966-1990.* Los Angeles: Higher Education Research Institute, UCLA.

Lenfant, C.
 1989 *Review of the National Institutes of Health Biomedical Research Training Programs.* Bethesda, Maryland: National Institutes of Health.

National Institutes of Health
 1992 Statement of work to the National Research Council. Mimeographed. March. Washington, DC.

CHAPTER TWO

APPROACHES TO THE ESTIMATION OF NATIONAL NEED

The vitality of the U.S. health research effort depends upon the availability of scientists who are personally committed to research, have mastered the theories and techniques of science, and can communicate their findings and assimilate new knowledge. In the biomedical and behavioral sciences, the nation's need for these research scientists is tied closely to problems of human health as well as opportunities for employment.

Previous National Research Council study committees have focused largely on education and employment in the biomedical and behavioral sciences and the role of the National Research Service Awards (NRSA) program in maintaining an adequate supply of well-trained scientists to maintain stability and efficiency in that system (IOM, 1985).[1] Given that approach, a substantial data base was developed over the years that permitted study committees to monitor trends in enrollment, degrees, employment patterns, and funding in certain of the fields. Analytic models of the training system were developed through surveys and special studies (see, for example, NRC, 1978). Supply-demand models were generated in the basic biomedical, behavioral, and clinical sciences to gauge the nation's need for scientists to maintain the balance in the education-employment system (see, for example, NRC, 1975, 1981, 1989).

With this report we mark a departure from the activities of previous NRC study committees. In recognition of the dynamic forces that create the demand for these highly skilled investigators, we present information on what we believe to be the key contextual variables that influence the size and quality of the research work force in the biomedical and behavioral sciences: the priority given by the nation to health research; recent major advances in the fundamental knowledge base, which works both to attract young people to scientific careers and to challenge them to master the latest scientific and technical developments; and changes in the demographic composition of the biomedical and behavioral science work force and recent employment experiences of its members. Our focus is the development of the science career and the role of the NRSA program in facilitating career development.

On the basis of our review of available information, we conclude that the demand for skilled research personnel in the biomedical and behavioral sciences continues to be strong. In the sections that follow, we identify what we believe to be the primary forces influencing the nation's need for skilled scientists in the next several years.

HEALTH RESEARCH AS A NATIONAL PRIORITY

The remarkable productivity of biomedical and behavioral research in the United States is largely a result of national patterns of investment in research and development (R&D). In the past decade alone, U.S. support for health research and development nearly tripled, growing from a total investment of $9.6 billion in 1982 to $28.1 billion in 1992 (NIH, 1993). This significant growth occurred when other research sectors experienced less dramatic support, often influenced by shifting priorities in federal budgets. The Defense Department's support for R&D, for example, grew rapidly in the early 1980s but peaked in 1986 and then declined. This change is reflected in the relative standing of the life sciences, engineering, and the physical sciences since 1980 (Figure 2-1).

The primary sponsors of health research in the United States are government agencies, industry (primarily the pharmaceutical industry), and private nonprofit organizations (including foundations, voluntary health agencies, and medical research organizations). (See Appendix Table F-1).

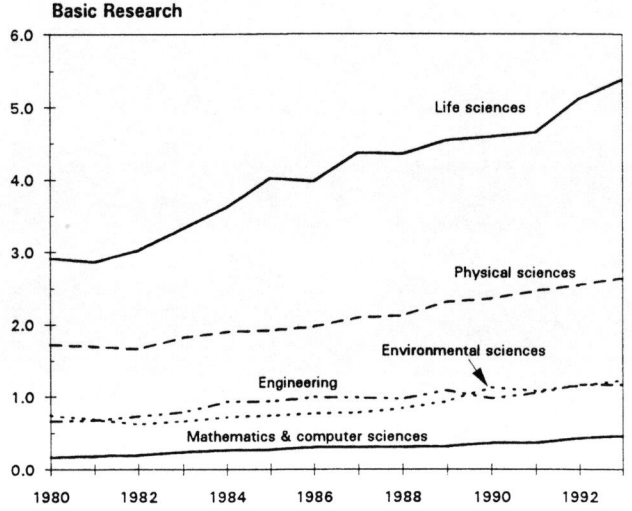

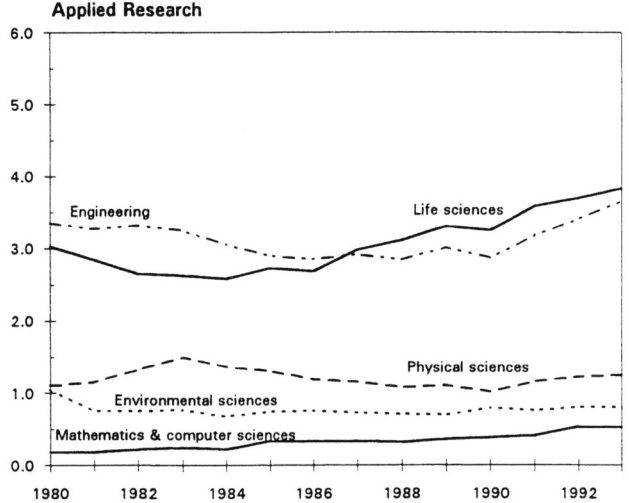

FIGURE 2-1 Federal obligations for research, by field (in billions of constant 1987 dollars). SOURCE: National Science Board, 1994.

Of the estimated $28 billion invested in health R&D in 1992, industry sponsored 45 percent, federal government agencies sponsored about 41 percent, and the nonprofit sector sponsored 4 percent.

Health research has also grown as a share of total national R&D, from about 13 percent in 1980 to 18 percent in 1992. As Figure 2-2 reveals, health R&D has grown as a fraction of total federal investment since 1984. However, health R&D has grown more rapidly as a share of all other national sources of support for research and development. In summary, although government budgets for health-related R&D have grown steadily over the past decade, other sectors have absorbed a greater share of the sponsorship of health research.

Patterns of Federal Support

The principal sponsor of government research in the health sciences in the United States is the National Institutes of Health (NIH). It is estimated that NIH will provide over 90 percent of the $10.9 billion of federal budget for health research and development in 1994. (See Appendix Table F-2). Research highlights from the fiscal 1994 health budget for the federal government reveal that the National Cancer Institute has the largest share of R&D funding within NIH ($2.08 billion in 1994), closely followed by the National Heart, Lung, and Blood Institute ($1.3 billion). These two institutes account for about one-third of the total NIH R&D budget. Although the administration proposed a 3.2 percent increase over fiscal 1993 levels for each of the 20 institutes and centers within NIH over fiscal 1993 levels, the U.S. Congress responded by increasing fiscal 1994 levels by over 5 percent. However, this rate of growth between fiscal years 1993 and 1994 represents a lower rate than that observed in the late 1980s which averaged about 8 percent a year. Components of the NIH budget in 1994 receiving the largest increases included:

- the Human Genome Center (23 percent),
- the National Library of Medicine (16 percent),
- the Fogarty International Center (10 percent),
- the National Institute for Allergy and Infectious Diseases (8 percent), and
- the Division of Research Resources (6 percent).

In summary, federal support for health research remains strong although the rate of growth has slowed somewhat in recent years (AAAS, 1992 and 1993). Coupled with current economic considerations (described below), it is uncertain, however, whether anticipated growth for health research will match that observed in the 1980s.

Trends in Industrial Support

In 1987 the pharmaceutical industry provided almost $5.4 billion and the biotechnology industry provided $1.4

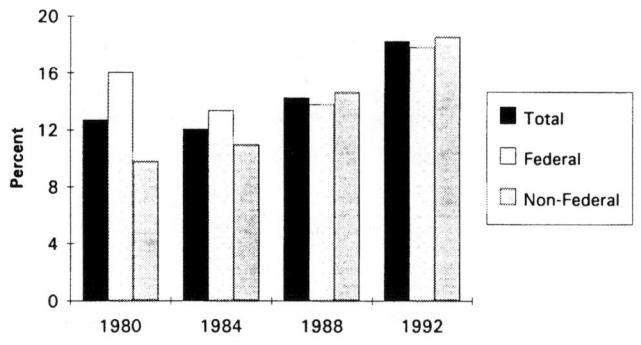

FIGURE 2-2 Funding of health R&D as a percentage of total R&D, by source. SOURCE: National Science Board, 1994.

billion for health R&D (OTA, 1988). The Pharmaceutical Manufacturers Association reported that the combined R&D expenditures of its member firms exceeded the total NIH budget in 1989. For several years industry has been viewed as the most rapidly growing sector of health R&D, but recent reports suggest that the growth of industrial investments in R&D may be leveling off, a result of corporate restructuring, slowdowns in productivity and economic growth, and other general economic conditions that inhibit industrial spending in activities that yield long-term payoffs rather than short-term gains (GUIRR, 1992).

Other Factors

In considering the future of health research in the United States during the next decade, it is important to realize that research progress may be in conflict with or limited by economic possibilities. These limiting factors include the impact of the budget deficit on federal spending, the economy, the unknown effects of health care reform, and the regulatory costs of research investments.

Impact of the Budget Deficit

Federal agencies throughout government are experiencing the costs and uncertain future associated with the mounting federal debt. As a result, efforts are underway throughout the Congress, the Office of Management and Budget, and elsewhere to cut unnecessary government spending and to reduce costs. In recent years there has been little annual increase in the research budgets for NIH after adjustment for inflation, apart from certain areas designated as priorities or major initiatives (such as the Human Genome Project).

The Economy

Industrial investments in health R&D were a major source of the growth experienced in this sector in the 1980s and compensated in part for reductions in federal growth patterns. However, slower economic growth and the lack of short-term market payoffs from prior research investments have raised questions as to whether significant increases in private sector funds allocated for health R&D can be sustained in the 1990s.

Unknown Effects of Health Care Reform

National interest in health care reform has raised many questions about the extent to which fundamental changes in physician-patient relations, co-payment arrangements, and the financing of medical and hospital care will affect the conduct and support of research. The emerging shift in emphasis from diagnosis and treatment of disease to the promotion of good health and the prevention of physical and mental disorders, for example, may deeply influence future research priorities in the health sector.

Regulatory Costs of Research Investments

The conduct of research has become part of a broader social and institutional context, and various social objectives or interests have generated a range of regulatory requirements and oversight mechanisms that carry additional costs. Issues such as the protection of human and animal research subjects, the protection of the environment (including the handling and disposal of toxic substances and the siting of hazardous facilities), and the need to ensure research integrity and fiscal accountability have raised new questions about the full costs of the research enterprise and the extent of the impact of hidden infrastructure costs on future research budgets.

ADVANCES IN RESEARCH

Another factor that plays a significant role in influencing the nation's need for biomedical and behavioral scientists is the growth of science. If we start, for example, from the work of the biologists and physicists who first gave us the structure of DNA 40 years ago, we can trace how that very fundamental discovery opened the field of molecular biology. The introduction of recombinant DNA technology was another milestone that gave us the tools to understand the gene.

In addition to being able to sequence genes rapidly, we can clone them and over-express proteins. The tools of physics and computer engineering have allowed the development of instrumentation that enables us to determine the structures of the macromolecules rapidly and accurately. This, in turn, has given rise to the field of structural biology.

The point is that basic science has given us the field of biotechnology, which now allows us to explore biology and medicine in more effective and exciting ways. It is this kind of serendipity that has caused science and medicine to advance.

The remarkable accomplishments of the biomedical and behavioral sciences are evident in the answers that are emerging to a host of disease problems—such as cancer, heart disease, acquired immune deficiency syndrome (AIDS)—and to the revolutionary advances in every branch of medicine. The success of fundamental research in the past few years can only be expected to accelerate the pace of discovery in the near future.

In the chapters that follow we identify the recent advances in the biomedical and behavioral sciences which we believe hold promise for attracting bright students to think creatively about the further development of science and its application to the health needs of the nation. Certain of these advances have been driven by significant investment in problems of national concern such as substance abuse or violence. However, other advances represent the next steps that will occur through the accretion of knowledge or the development of new technologies that make exploration of new problems possible.

MARKETPLACE REQUIREMENTS

The first report of the National Research Council on the subject of national needs for biomedical and behavioral research personnel was issued in June 1975, only 4 months after the Council accepted the task proposed under the NRSA Act of 1974. Because of time constraints, the first study committee devoted its initial report to a description of the organization of the study, an outline of the issues involved, and a presentation of the limited data available at that time. Each subsequent study committee updated or enlarged the scope of topics addressed in prior reports and included some new issues.

In organizing the first study, the 1975 NRC committee divided the biomedical and behavioral fields into four areas: (1) basic biomedical sciences, (2) behavioral sciences, (3) clinical sciences, and (4) health services research.[2] A panel of experts was formed to assist the committee in each area, and an additional panel was created to guide the data collection and analyses.

It was recognized very early in that study that the legislative request to specify the nation's personnel needs in the fields of biomedical and behavioral research would be impeded by the difficult problems of definition and classification. An attempt was made, therefore, in the first report to define each of the four broad areas in terms of the disciplinary fields included in them. These initial definitions were revised in subsequent reports, but the problems of taxonomy and determining need at the disciplinary level continue to be among the most intractable ones facing every committee. The major problem, as pointed out in the 1975 report, is that the boundaries between disciplines are difficult to draw. This problem is compounded by the adaptability of biomedical and behavioral scientists and their capacity for mobility within and across fields. This is especially true for transfers from more fundamental to more applied fields and for transfers facilitated by postdoctoral training. Lastly, there is the difficulty of predicting major scientific developments and their potential impact on personnel requirements. In view of these considerations, the recommendations of previous NRC committees have been directed almost exclusively to broad areas rather than to disciplinary subgroups.

The 1975 study committee commented on the weaknesses inherent in all available manpower models when it first began to assess methods of projecting the labor market. That committee observed that supply depended largely on such intangibles as students' perceptions of the prestige, prospects, and value of various careers. Demand, in turn, responded to another set of factors—future economic conditions, levels of federal funding, and evolving research priorities—all of which could change abruptly and unpredictably.

Nonetheless, during the 1970s the study committees used supply and demand models to assess the relationships between the production of Ph.D. scientists, the need for research personnel in universities, and the commitment of funds from various sources to R&D in the basic biomedical and behavioral sciences. However, in 1981, another study committee reported that significant problems existed in making projections from available data. First, the latest data were not always available for use in the model. Second, changes within specific disciplinary subgroups were not measurable (only the number of people in broadly defined categories was known), thus limiting the utility of the analysis to estimate "national need". Third, assumptions about job mobility were flawed: it had been assumed in the past that turnover in the pool of career scientists in U.S. colleges and universities was principally due to retirement and death. More recent assessments indicated this was not the case, that there was a substantial amount of switching between academic and nonacademic positions.

During the mid-1980s, succeeding committees continued to use supply and demand models, introducing certain improvements.

In particular, the 1989 study committee enhanced earlier analyses of the labor market for biomedical and behavioral scientists in several ways:

• expanded the labor market analysis to include industry, government, hospital, and other nonacademic sources of labor demand;
• developed separate projections for the labor market in general and for scientists working in R&D or management of R&D;
• included a demographic-economic model for estimating scientist attrition due to death, retirement, and net occupational movement;
• brought labor supply into the labor market assessment; and
• projected labor market variables to the year 2000.

Another improvement over the years has involved the disaggregation of disciplinary fields for purposes of analysis, specifically in the behavioral sciences. In the earliest reports, study committees presented the labor market outlook for Ph.D.s in the behavioral sciences as a whole, without distinguishing between the subfields of psychology, sociology, anthropology, and speech and hearing science. In 1978 the study committee, realizing that the analysis of the labor market was hindered by treating the behavioral sciences as a single entity, separated the data into clinical (clinical psychology, counseling and guidance, and school psychology) and nonclinical fields. This disaggregation enabled the identification of divergent market trends within the behavioral sciences.

Status Report on the Work of the Panel on Estimation Procedures

The Panel on Estimation Procedures was asked by the present committee to analyze models of supply and demand used by previous NRC study committees. The panel decided to analyze the model used by the 1989 committee[3] and presented their results to the committee in September 1993.[4] While the Panel on Estimation Procedures plans to prepare a separate report summarizing its deliberations, a brief overview of their work follows.

1989 NRC Model

The 1989 NRC model includes a demographic model that projects the numbers and characteristics (primarily age) of the supply of scientists in the work force and a curve-fitting procedure that forecasts the demand for scientists by sector (NRC, 1989). The panel concluded that all components were found wanting. The demographic projection model confuses age and cohort effects; it could only project accurately in a steady state, when no projections are needed. It has no mechanism for projecting new entrants into the labor market. The demand model restates earlier ad hoc committee models and assumes, falsely, that ratios of students to teachers and of scientists to research dollars are fixed and do not change as economic and technical conditions change. Finally, the treatment of equilibrium implicitly (and incorrectly) assumes that today's excess demand or supply has no effect on tomorrow's market. The panel recommended to the committee that the 1989 NRC committee model "should not be the base the committee uses in developing its recommendations".[5] Instead, the panel recommended the exploration of alternative approaches to estimating supply and forecasting demand.

New Techniques for Estimating National Needs

The panel explored the feasibility of developing projections of supply through demographic techniques. (Preliminary work using this technique may be found in the next two chapters of this report.) These techniques begin by listing the characteristics of a given population (e.g., age, sector of employment, and employment status) and projects changes in the population based on the life history of members of the population. These multistate life tables (Keyfitz, 1985) are useful for answering the questions, What will be the characteristics of the labor force in 5 years? How long do workers remain in a particular job?

The answer is generated through a series of statistical calculations by making assumptions about both the rates of transition of individuals from state to state (employed to retired, for example) and the rates of new entrants to the system. The panel expects to continue its work in this area in 1994 and will prepare a final report containing recommendations for the further development of forecasting activities along these lines.[6]

The panel is skeptical, furthermore, about the possibility of generating useful forecasts of demand and gave two reasons for doubting that useful long run forecasts of demand could be made. First, the conceptual basis of current forecasting models is questionable. Demand forecasts in the tradition of previous reports assume the existence of some fixed function relating scientists needed to students and research dollars. They assume, in the simplest case, that student-faculty ratios are fixed and that the number of dollars required to support one scientist is fixed. This is, on its face, silly. Student-faculty ratios depend on market conditions and historically have varied considerably. The relationship between research dollars and Ph.D. scientists employed depends on the cost and capability of machinery and the relative costs of technicians, postdoctoral appointees, and Ph.D.s, all of which are absent from these models. More complex models assume complicated (and completely ad hoc) relationships between students and faculty and scientists and dollars spent on research; they continue to assume that these relationships are fixed and independent of economic and technological conditions.

The second reason the panel is skeptical about the possibility of generating useful long-term demand forecasts is simpler. To be of use, the forecasts must predict accurately a long period in the future. Those who enter graduate school today will be beginning their independent research careers 6 to 8 (or more) years from now. The predictions of forecasting models become imprecise when projected 5 to 10 years. Models that previous committees developed are a case in point.

The panel believes it is both feasible and useful to develop short-term indicators of demand. Monitoring these will provide information about current employment conditions. This information is of limited use to policymakers because today's decisions about fellowships and traineeships primarily affect the scientific labor market a decade from now. Nonetheless, it is useful for two reasons. First, it may suggest ways in which current policy should be changed. For example, if a field has an extremely tight labor market, a shift from post- to pre-doctoral support may be appropriate.

Second, current market conditions are an important influence on young people's decisions to become research scientists.

Summary

The Panel on Estimation Procedures has conducted a series of preliminary assessments of available models of supply and demand and has concluded that they are of no utility to this present effort to establish the nation's overall need for biomedical and behavioral research personnel. The committee has accepted this conclusion and suspended use of mathematical models of supply and demand with this study.

The panel has concluded that new approaches are needed to project the supply of these researchers and to estimate demand. The panel has suggested the use of multistate life tables for assessing changes in the composition of the labor force over time. The committee has accepted this suggestion. The panel has also offered sample short-term indicators of demand, some of which are included in this report.[7]

The committee believes that the panel has made a significant contribution to the process of establishing overall need for biomedical and behavioral scientists by demonstrating the potential value of techniques that monitor changes in the supply of scientists and the value of short-term indicators of demand over previous mathematical approaches to estimating supply and demand adopted by the NRC. Although the product of their work, found in the next two chapters, must be viewed as preliminary, it already shows promise as a policy tool in human resource studies.

NOTES

1. See, also, Appendix A for a brief overview of the key features of previous NRC reports in this area.

2. Legislative reform in the 1970s resulted in the inclusion of nursing research personnel around 1978. Oral health research personnel (included by that name in this present study) represented an outgrowth of the clinical sciences. Research training needs in that area were addressed separately as "dental research training" in the late 1970s.

3. The Panel reviewed earlier models developed by previous NRC committees but concluded that their attention to "academic employment opportunities" restricted their utility to present market concerns, and removed them from serious consideration for further use by the National Research Council.

4. M. Rothschild, report to the Committee on National Needs for Biomedical and Behavioral Research Personnel, September 11, 1993.

5. M. Rothschild, ibid. Rothschild continues: "Tinkering with the model while retaining its basic structure will not make it a useful tool."

6. Since future Ph.D.s are produced only by those prior Ph.D.s who entered academia, projections of the supply of future Ph.D.s must separate these two components of doctoral supply. Simple projections of future supply based on the current entire doctoral population are misleading.

7. The committee restricted the application of these new techniques to the market outlook in the basic biomedical and behavioral sciences. This was due to the fact that data requirements made detailed analyses possible only in those areas. To the extent data were available for a similar set of variables, they were included in other chapters. However, the multistate life table method was restricted to work in chapters 3 and 4 owing to the lack of information of sufficient detail to permit the application of that method in the other areas.

REFERENCES

American Association for the Advancement of Science
    1992   *Congressional Action on Research and Development in the FY 1993 Budget.* Washington, D.C.: AAAS.
    1993   *Congressional Action on Research and Development in the FY 1994 Budget.* Washington, D.C.: AAAS.

Government-University-Industry Research Roundtable
    1992   *Fateful Choices: The Future of the U.S. Academic Research Enterprise.* Washington, D.C.: GUIRR.

Keyfitz, N.
    1985   *Applied Mathematical Demography.* 2nd Ed. New York: Springer-Verlag.

Institute of Medicine
    1985   *Personnel Needs and Training for Biomedical and Behavioral Research.* Washington, D.C.: National Academy Press.

National Institutes of Health
    1993   *NIH Databook 1993.* Publication No. 93-1261. September. Bethesda, MD: National Institutes of Health.

National Research Council
    1975   *Personnel Needs and Training for Biomedical and Behavioral Research.* Washington, D.C.: National Academy Press.
    1978   *Personnel Needs and Training for Biomedical and Behavioral Research.* Washington, D.C.: National Academy Press.
    1981   *Personnel Needs and Training for Biomedical and Behavioral Research.* Washington, D.C.: National Academy Press.
    1989   *Biomedical and Behavioral Research Scientists: Their Training and Supply.* 3 Volumes. Washington, D.C.: National Academy Press.

National Science Board
    1994   *Science and Engineering Indicators - 1993.* Washington, D.C.: National Science Foundation.

Office of Technology Assessment
    1988   *New Developments in Biotechnology.* Washington, D.C.: Office of Technology Assessment.

CHAPTER THREE

BASIC BIOMEDICAL SCIENCES PERSONNEL

Exciting developments in the basic biomedical sciences continue to attract talented individuals to research careers. Although the National Institutes of Health (NIH) provide both extramural and intramural support to advance research and maintain a pool of skilled scientists, it is the National Research Service Awards (NRSA) program that provides the most promising young scientists with the funds they need to complete their training while pursuing research topics of interest to them and the nation.

Like previous NRC committees formed to address the future direction of the NRSA program, we have considered the appropriate level and mix of predoctoral and postdoctoral support in the basic biomedical sciences given advances in basic research and changing employment patterns for scientists in component fields (see Appendix B for a field taxonomy). This has been a challenging task for three reasons. First, changes at the NIH may well favor a shift toward more basic research in the health sciences. However, unless resources are made available to basic biomedical scientists to pursue those new directions, the connection between training and research will be broken. The continued success of the NRSA program depends on its ability to attract highly qualified and promising students to enter training and pursue careers in research. The interest of potential trainees in such a career and their ability to pursue it depends, in turn, on continued federal commitment to support health-related research as an important national need.

A second, possibly related, challenge we confronted involves the interpretation of the current and future market for basic biomedical scientists. We realize much has been written recently about the difficulties some young biomedical scientists have encountered in locating positions in research settings and/or securing research support. (Indeed, a recent report of the National Research Council's Commission on Life Sciences [NRC, 1994] addressed the topic of obtaining NIH support.) We believe that for some young scientists the market has become sluggish. We cannot help but observe, however, that the unemployment rate of basic biomedical scientists is estimated to be about 2 percent or less and has not changed significantly in the past two decades. We attribute this finding to the fact that not all Ph.D.-level scientists pursue careers in academic research settings; some work in government, industry, or schools. Almost all are essential, however, to the support of the infrastructure of the nation's research and training enterprise in the basic biomedical sciences.

Our Panel on Estimation Procedures has persuaded us that the mathematical models of supply and demand found in previous reports on the NRSA program should be abandoned in favor of an analysis of the supply of scientists and a separate look at selected indicators of market conditions. As noted in the previous chapter (Chapter 2) mathematical models of supply and demand have a number of deficiencies, among them a lag in information about the most recent employment prospects. While the new techniques developed by the panel have not solved the problem of having up-to-date market information, the analyses inherent in the indicators of market conditions coupled with the use of multistate life table analyses of changes in supply represent at least a partial solution. These analyses do not offer a specific assessment of future demand. Because of this, the committee chose to develop alternative assumptions about the growth of future demand and to examine the implications of these assumptions for the number of degrees in biomedical sciences that would be required to meet these assumed rates of growth. The product of that assessment may be found in the pages that follow.

On the basis of that analysis and our subsequent deliberations, we conclude that the national biomedical effort continues to benefit from the steady addition of men and women (including minorities) to the basic biomedical sciences work force. They appear to be employed productively, although not all may be working in research laboratories. The best predictions for economic activity and research and development (R&D) funding in the near future suggest, however, that demand for basic biomedical scientists will grow slowly. Our primary concern at this time, therefore, is to maintain the supply of highly skilled scientists required to

- keep our nation in a lead position in all areas of basic and applied biomedical sciences,
- respond rapidly and effectively in combating new problems in human health and disease, and
- ensure the efficient transfer of new knowledge and technology into developing areas of clinical promise and industrial opportunity.

Meeting the national need for highly qualified and productive biomedical scientists depends on attracting sufficient numbers of the best and brightest high school and college students into scientific careers at the graduate level, which in turn depends on facilitating access of all qualified students to this education path. The NRSA program clearly has an important role to play in that effort.

In considering the future role of the NRSA program in meeting national needs for bioscientists, we confronted our third challenge: the increasing attractiveness of stipends for research training from sources other than the NRSA program. We learned from participants at the public hearing (May 1993, Appendix C) that the NRSA program generally remains effective in recruiting individuals into the research training path and launching them into research careers. However, for reasons largely related to years of stagnant growth in stipend support, the NRSA is no longer competitive with other mechanisms of training support, which have higher stipends and more flexibility. The committee considered these issues and concluded that high priority must be given to restoring appropriate stipend support through the NRSA program even at the expense of overall growth in the total number of awards in the basic biomedical sciences over the next few years. Thus, our recommendations for the future direction of the NRSA program reflect our deep conviction that the NRSA program must continue to play a significant role in the national biomedical research effort and that this will require prompt attention to issues of stipend size and flexibility. In the pages that follow we shall address each of these issues.

ADVANCES IN RESEARCH

Innovations in basic science and in technology are inextricably intertwined and inseparable. Advances in basic science lead to development of new technologies that, on the one hand, give rise to new therapies and industrial applications, and on the other hand, give rise to fresh advances in science from which, in turn, emerge additional new technologies. Thus, advances of the 1960s and early 1970s in biochemistry, microbiology, and genetics provided the information required for development of recombinant DNA technology, which has, in turn, provided the basis for revolutionary new insights in many fields of biology and medicine and has given rise to novel diagnostic and therapeutic modalities and to an entirely new biotechnology industry. More recently, the discovery of extremely thermophilic bacteria that live in hot springs and deep ocean vents and study of their biochemistry, together with other advances in molecular biology and recombinant DNA technology, has led to a new refinement in molecular biological analyses, the polymerase chain reaction (PCR). PCR has proven to be an extraordinarily powerful tool in basic biomedical and clinical research and has important applications in areas of medical biotechnology, such as clinical diagnostics. Indeed, it is noteworthy that Dr. Kary Mullis shared the 1993 Nobel Prize in Chemistry for his work in development of PCR.

A second new technology with major applications in clinical medicine and industry, as well as basic biomedical research, arose from immunologists' need to understand the nature of the immune response and the mechanisms controlling the formation of antibodies. The so-called monoclonal antibody technique provides a method of exquisite specificity and sensitivity for identifying and purifying any molecule that is capable of eliciting an immune antibody response. Current medical applications include, for example, rapid and precise identification and analysis of pathogenic microorganisms and tumor cells and purification of specific types of immune cells for diagnostic and therapeutic purposes.

These and other innovative technologies, in conjunction with advances in microchemistry, instrumentation, and, most notably, computer science, have fueled a continuing explosion of understanding in many fields of basic biomedical sciences. These include mechanisms of regulation of gene expression in growth, differentiation, and development; biochemical mechanisms regulating normal and abnormal cell growth and multiplication; mechanisms whereby the immune system recognizes, processes, and responds to antigenic stimuli; protein structure at atomic resolution, the relationship of structure to specific protein function, and the principles of protein design; mechanisms of nervous system development and function, including molecular bases of learning and memory; elucidation of the human genome and its expression; and development of new food crops to feed the world's population.

Advances over the past decade in areas of basic biomedical science such as those cited above have profound implications for understanding the genesis of major human diseases and for future development of effective means of prevention, therapy, and cure. Efforts to map the human genome and identify mutant genes responsible for heritable diseases are progressing rapidly under the aegis of the Human Genome Project and form the basis of the newly emerging field of gene therapy. It is worth noting that studies on the human genome depend heavily on technologies which derive from concurrent work on mouse genetics and embryology, such as construction of transgenic mice. These technologies have also been important for discovery and functional analysis of so-called oncogenes and tumor suppressor genes. The protein products of those genes normally serve as important regulators of gene expression or cellular growth and multiplication but can, through mutation or aberrant expression, trigger unregulated cancerous growth of cells. The rapidly increasing understanding of the complicated biochemical and genetic means by which control of cell growth and multiplication is achieved—and lost—are providing major new insights into the root causes of cancer and possible strategies for prevention and therapy. During the past decade, structural biology (specifically, determination of three-dimensional macromolecular structure at atomic resolution) has undergone a renaissance with the advent of recombinant DNA-based methods for production of large amounts of pure proteins and nucleic acids and concomitant improvements in instrumentation and computational methods. This information is crucial, not only for understanding functional interactions of normal and abnormal proteins, but also for determining how drugs interact with their target proteins and for rational design of new agents with improved therapeutic properties.

ASSESSMENT OF THE CURRENT MARKET FOR BASIC BIOMEDICAL SCIENTISTS

Employment conditions for biomedical scientists were relatively robust throughout the 1980s (Figure 3-1). In response to expanding opportunities in health research, the basic biomedical work force grew dramatically, rising from roughly 64,000 Ph.D.s in 1981 to nearly 92,000 in 1991. This 44 percent growth is about twice that of the total science and engineering work force and quadruple the rate of employment growth of the total U.S. work force.[1]

Accompanying this dramatic work force growth were substantial changes in its composition. Among the notable changes were the growing prominence of females and Asians[2] and the declining prominence of native-born male citizens. Accompanying these changes in work force characteristics were changes in the nature of employment opportunities. Academic employment declined as a relative share of total employment as industrial employment grew.

As Figure 3-2 suggests, about 23 percent of the basic biomedical science work force were women in 1991, up from 17 percent in 1981. Almost half of the women in the

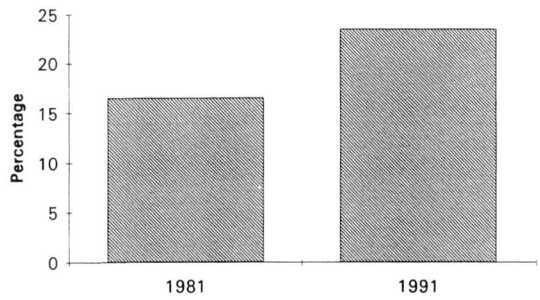

FIGURE 3-2 Fraction of the U.S. biomedical science work force who are women, 1981 and 1991. See Appendix Table F-3.

1991 work force were younger than 40, compared with roughly 38 per cent of the men.[3]

The biomedical sciences work force has also become more racially diverse over the years, but progress has been slow. In 1991 nearly 12 percent of the employed biomedical science Ph.D.s represented individuals from a racial minority group (Table 3-1). In 1979 these minorities represented about 8 percent of the biomedical work force. Most of the growth occurred for Asians. Progress in ethnic diversity is less dramatic. In 1991 Hispanics represented about 2 percent of the biomedical scientists. In 1979 the comparable statistic was roughly 1 percent.

The age distribution of the work force is an important early-warning indicator of future replacement needs. Despite its rapid growth over the past decade, the biomedical work force is aging (Figure 3-3). The median age has risen slowly from 39 to 42 years. Based on these projections, the 1989 study concluded that annual replacement needs would increase by 28 percent between 1987 and 1995, from 5,086 to 6,543 as a result of replacement demand (NRC, 1989).

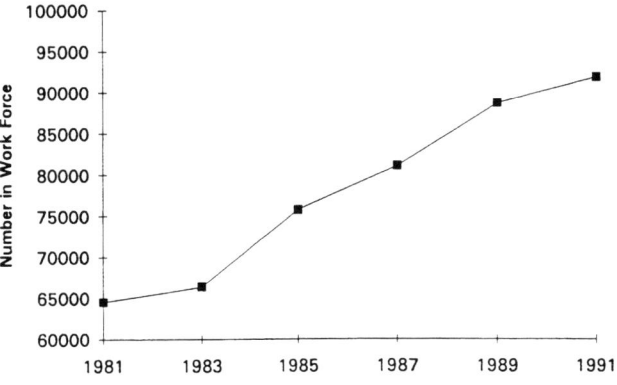

FIGURE 3-1 U.S. biomedical science workforce, 1981-1991. See Appendix Table F-3.

TABLE 3-1 Racial/Ethnic Composition of the Employed Biomedical Ph.D.s: 1981 and 1991

	1981[a]		1991[b]	
	Number	Percent	Number	Percent
Race				
TOTAL	58,264	100.0	85,275	100.0
White	53,005	91.0	75,830	88.9
Black	722	1.2	1656	1.9
Asian/Pacific Islander	4,438	7.6	7,583	8.9
Other (Incl. Native American)	99	0.2	206	0.2
Ethnicity				
TOTAL	56,950	100.0	84,803	100.0
Hispanic	818	1.4	1,263	1.5
Non-Hispanic	56,132	98.6	83,540	98.5

[a] For those who responded in 1981. Race nonresponse was 182 in 1981 and ethnic nonresponse was 1,496.
[b] For those who responded in 1991. Race nonresponse was 321 in 1991 and ethnic nonresponse was 793.

NOTE: Employed biomedical Ph.D.s are those with a biomedical Ph.D., regardless of employment field. Estimates are subject to sampling error. Comparisons between 1991 estimates and those of earlier years should be made with caution due to changes in survey methodology. Prior to 1991, the SDR collected data by mail methods only. In 1991, the survey had both a mail component and a telephone follow-up component. In this table, 1991 estimates are based on "mail-only" data to maintain greater comparability with earlier years.

SOURCE: NRC, Survey of Doctorate Recipients. (Biennial)

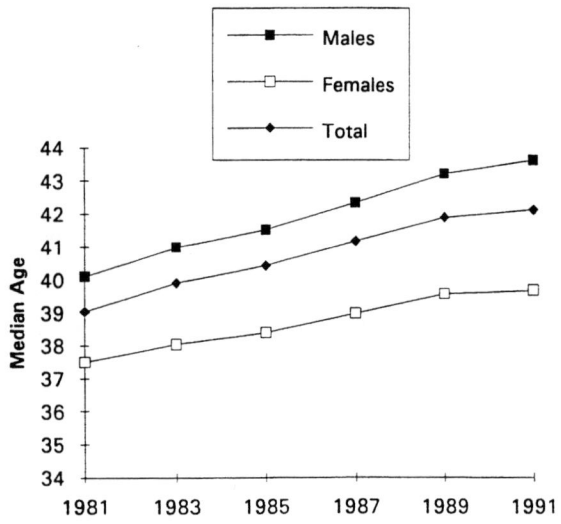

FIGURE 3-3 Median age of U.S. biomedical science work force gender, 1981-1991. See Appendix Table F-3.

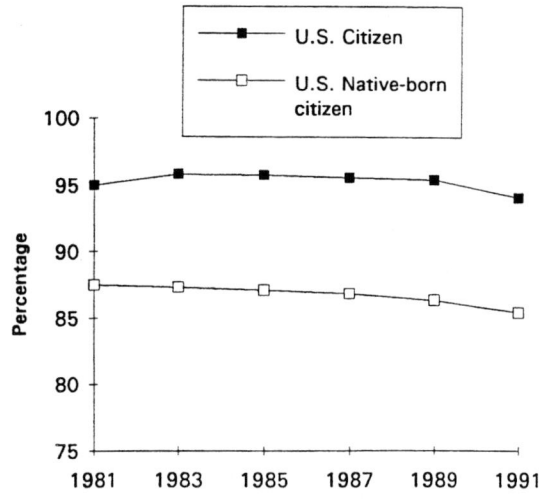

FIGURE 3-4 Citizenship status of employed biomedical science Ph.D.s, 1981-1991. See Appendix Table F-4.

Almost 95 percent of employed biomedical science Ph.D.s were U.S. citizens in 1991 (Figure 3-4). However, between 1981 and 1991, native-born U.S. citizens became a smaller proportion of the total U.S. biomedical work force. That is, in 1981 native-born U.S. citizens represented 87.5 percent of the biomedical work force, compared with 85.4 percent in 1991. While weak, this declining trend reflects the more general phenomenon in science and engineering as a whole, wherein U.S. native-born citizens represented 86.4 percent of the Ph.D. work force in 1981 but only 82.7 percent ten years later (NSF, 1990).

Individuals holding postdoctoral appointments are an important component of employment in the biomedical field. Proportionately, more women than men held postdoctoral appointments.[4] In part, the difference is because the likelihood of holding a postdoctoral position var-

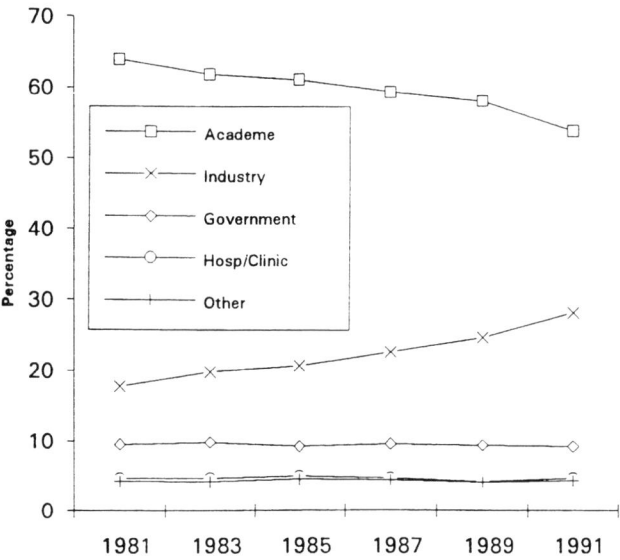

FIGURE 3-5 Employment sector of the U.S. biomedical science work force, 1981-1991. See Appendix Table F-5.

ies inversely with career age, and women, as more recent participants in biomedical science, tend to be younger.

Opportunities for employment in the academic sector have grown more slowly than have opportunities for employment in nontraditional settings (Figure 3-5). As a result, only 54 percent of the biomedical science work force were employed in academia in 1991 in contrast to two-thirds in 1981. This trend reflects the changes in academic employment prospects experienced by those in many other fields. Offsetting this trend, however, has been the dramatic rise in employment opportunities in industry. This sector accounted for almost 28 percent of 1991 employment, up from almost 17 percent in 1981. The proportion of the basic biomedical workforce employed in other sec-

tors (such as government or in hospitals and clinics) remained about the same during that period.

Degree Production and Career Patterns

The major source of new biomedical science talent has been our nation's university system. It is not the only source of talent, however. Some jobs are filled by immigrants who received their degrees in other countries. Furthermore, some recipients of biomedical Ph.D.s are employed in other fields, and some biomedical science jobs are filled by workers with degrees in other fields.

Degree Production

The most readily available source of information about patterns of degree production is the Doctorate Records File,[5] which describes degree production from U.S. universities; the committee summarizes this information below.

A declining trend in degree production occurred between

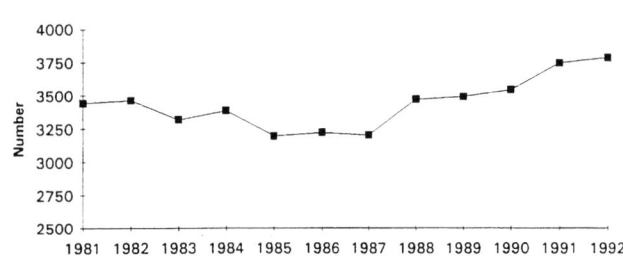

FIGURE 3-6 Biomedical science Ph.D. production, 1981-1992. NOTE: Data limited to U.S. citizens and permanent residents. See Appendix Table F-6.

1981 and 1985 and was followed by a relatively stronger upward trend between 1987 and 1992. The annual number of degrees produced in the biomedical sciences rose by 10 percent over the entire period, from about 3,400 to almost 3,800 (Figure 3-6). This rate of increase was notably slower than the comparable rate of 31 percent for doctorates in all fields of science and engineering (Ries and Thurgood, 1993).

There were notable changes in the characteristics of the degree recipients in the biomedical sciences: an increasing fraction were female and a smaller fraction were U.S. citizens. The average age of recipients increased.

Significant progress has been made in achieving gender diversity. The number of degrees granted to women increased between 1981 and 1992 by almost 60 percent (from roughly 1,000 to about 1,600). In 1981 women represented 29 percent of the degrees produced in biomedical sciences; by 1992 they received 43 percent (Figure 3-7).

Little progress has been made with respect to race and

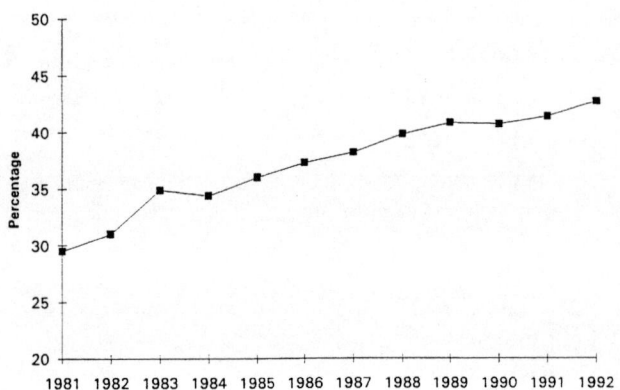

FIGURE 3-7 Fraction of biomedical science Ph.D. degrees earned each year by women, 1981-1992. NOTE: Data limited to U.S. citizens and permanent residents. See Appendix Table F-6.

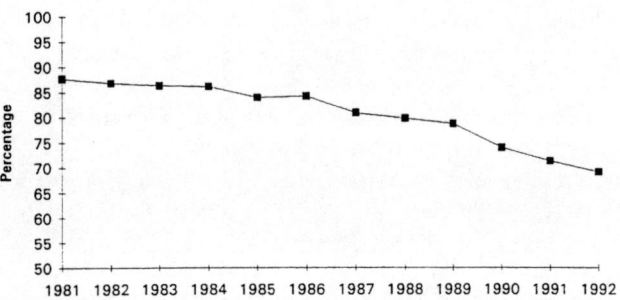

FIGURE 3-8 Fraction of biomedical science Ph.D. degrees earned each year by U.S. citizens, 1981-1992. See Appendix Table F-7.

ethnic diversity, however (Table 3-2). When analyses are restricted to degree recipients who are U.S. citizens or permanent residents, we find that whites constituted about 91 percent of 1981 degree production; in 1992 they represented just over 86 percent. Roughly half of this small decline can be accounted for by the growth in the number of degree recipients of Asian origin. The share of degrees awarded to Asians rose from 5.3 in 1981 to 8.4 percent in 1992.

There has also been a dramatic change in the citizenship status of biomedical degree recipients. The percentage who were U.S. citizens declined from 88 percent in 1981 to 69 percent in 1992 (Figure 3-8). Similar changes are occurring in other fields, particularly in the physical sciences and engineering (NSF, 1990). This change may ultimately be reflected in the citizenship characteristics of the biomedical work force.[6]

Career Patterns

Given the objective of the NRSA awards—to produce research scientists—it is useful to have some notion of the number of years over the course of a career that these scientists remain engaged in R&D. The effectiveness of the program will vary with this number. The Survey of Doctoral Recipients — a longitudinal survey that tracks doctorates in the sciences, engineering and humanities biennially — provides useful information on employment patterns, including postdoctoral work. This survey has the potential for illuminating career patterns of biomedical scientists. Thus, the Panel on Estimation Procedures will examine more closely the feasibility of estimating such patterns.

Market Conditions

This section presents short-term indicators of market conditions: unemployment and underemployment rates, postdoctoral appointments, postgraduation commitments of

TABLE 3-2 Biomedical Ph.D. Production Over Time, by Race and Ethnicity

		1981	1982	1983	1984	1985	1986	1987	1988	1989	1990	1991	1992
Total	N	3293	3359	3243	3294	3126	3162	3119	3406	3429	3482	3684	3728
White	%	91.5	91.1	90.4	90.3	89.9	89.5	88.9	89.5	88.7	88.5	86.3	86.3
Black		1.9	1.8	1.6	1.9	2.0	1.6	2.3	1.9	2.2	2.0	2.3	2.1
Hispanic		1.2	1.5	1.4	1.5	1.8	2.1	2.2	2.4	2.3	2.6	2.7	2.7
Asian		5.3	5.4	6.3	6.0	5.9	6.2	6.2	6.0	6.5	6.7	8.3	8.4
Native American		0.2	0.2	0.2	0.3	0.4	0.6	0.3	0.2	0.3	0.1	0.3	0.5

NOTE: Cases with missing data are excluded. Data limited to U.S. citizens and permanent residents.

SOURCE: NRC, Survey of Earned Doctorates. (Annual)

new doctorates, and relative salaries.[7] No strong trends have been discerned, although changes in postgraduation commitments and starting salaries suggest that the demand for basic biomedical scientists has been growing more slowly than in earlier years.

Unemployment and Underemployment

The most commonly used short-term indicator of labor market conditions is the unemployment rate. In labor markets for highly skilled workers, however, the unemployment rate is not as meaningful as an indicator of market conditions. This is because such workers are able to find jobs even in times of weak demand. Thus, the issue is not whether the worker has a job, but whether the job is fully utilizing the worker's skills. For this reason, the committee has also compiled information on underemployment which

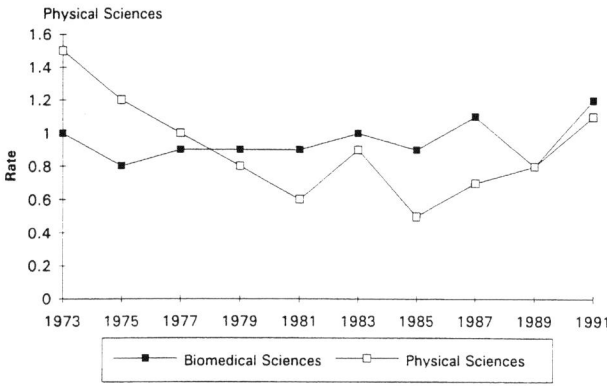

FIGURE 3-10 Underemployment rates of biomedical and physical sciences Ph.D.s, 1973-1991. See Appendix Table F-9.

Postdoctorates

The number of new Ph.D.s with postdoctoral appointments can also reflect labor market conditions.[8] One of the functions of postdoctoral appointments, for example, has been to provide interim positions for new researchers. Given this as one of many functions of postdoctoral appointments, this number can be expected to rise when demand is weak.

Figure 3-11 summarizes information on such appointments for new biomedical researchers (i.e., those who received their degrees 4-5 years earlier) for the period 1973-1991. The data show that the fraction of these researchers who are postdoctorates rose dramatically in the 1970s. This strong trend was followed by a smaller, more erratic pattern in the 1980s.

These trends, on inspection, do not support the notion of a weakening demand in the biomedical fields in the 1980s. The unusual pattern observed in the 1980s suggests that other factors may have influenced the fraction of recent bio-

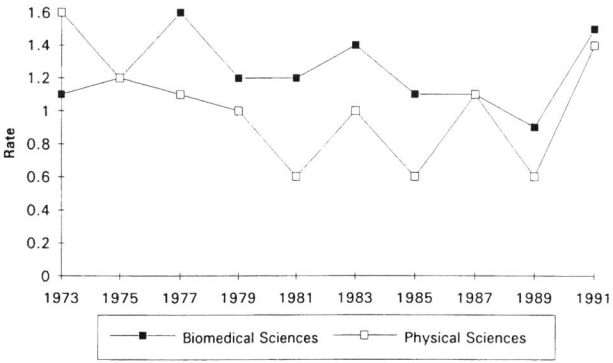

FIGURE 3-9 Unemployment rates of biomedical and physical sciences Ph.D.s, 1973-1991. See Appendix Table F-8.

is defined to include workers who are working part time but would prefer full-time jobs and workers who have jobs that are outside of science and engineering and who indicate they took these jobs because they could not find work in science and engineering. Figures 3-9 and 3-10 summarize these rates. Because concern has been expressed recently about the weak state of demand in the physical sciences, comparable rates for physical scientists are included so that the reader can assess the relative status of biomedical science labor markets as gauged by this indicator.

Several conclusions emerge. First, as noted above, unemployment is not a serious problem. Rates of unemployment and underemployment generally hover around 1 percent in each of the fields examined. The data in Figure 3-9 contrast strikingly with the rate for the entire U.S. work force, which has ranged between 4.9 and 6.7 percent during this period (Office of the President of the United States, 1993).

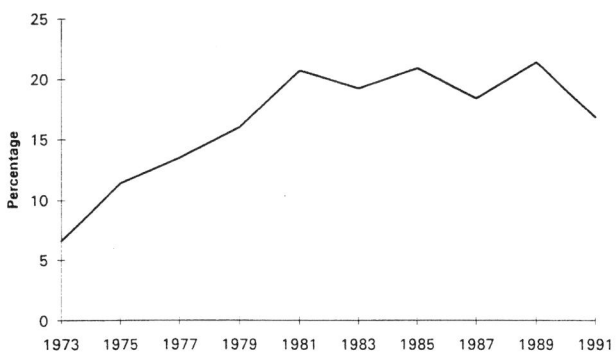

FIGURE 3-11 Fraction of biomedical science Ph.D.s at career age 4-5 on postdoctoral appointments, 1973-1991. See Appendix Table F-10.

medical Ph.D.s holding postdoctoral appointments in that period. For example, the observed trends may reflect variations in the availability of funding for postdoctorates, with increasing support in the 1970s and increasing constraint in the 1980s.

Postgraduation Commitments

The postgraduation plans of new doctorates may also reflect market conditions. In particular, the percentage of new doctorates who indicate that they have definite commitments at the time they are completing their requirements for the degree can reflect the strength of demand. When demand is weak this percentage will fall; when demand is strong this percentage will rise.

Figure 3-12 summarizes these plans for the period 1975 to 1992. To provide a comparative base, similar information is provided for degree recipients in the physical sci-

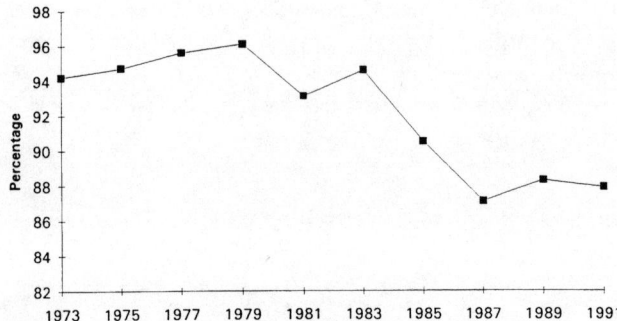

FIGURE 3-13 Salaries of biomedical science Ph.D.s (age 30-34) who currently hold full-time employment positions (excluding postdoctoral positions) as a percentage of comparable salaries for all scientists and engineers, 1973-1991. See Appendix Table F-12.

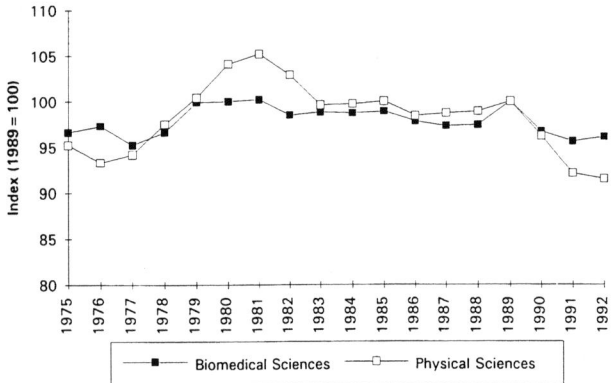

FIGURE 3-12 Fraction of new biomedical and physical sciences Ph.D.s with definite commitments, 1975-1992. See Appendix Table F-11.

ences, which are thought to be suffering currently from weak demand. The data show a notable declining trend in this percentage for each of these fields beginning in 1989, but the trend is more pronounced in the physical sciences.

Starting Salaries

"Starting salaries" are defined as the median salaries of doctorates, age 30-34, who currently hold full-time employment positions (excluding postdoctoral positions). Information regarding the starting salaries of biomedical doctorates relative to comparable salaries for all science and engineering doctorates is presented in Figure 3-13. Since 1983, salaries for these scientists in fields other than the basic biomedical sciences taken as a whole have been growing relatively faster than salaries for basic biomedical sci-

entists. This suggests that relative demand has been growing more slowly in the biomedical sciences than in other fields of science or engineering combined.[9]

OUTLOOK FOR BASIC BIOMEDICAL SCIENTISTS

The labor market for biomedical scientists defines one dimension of need. Job openings are generated by deaths, retirements, and other types of separation from the biomedical work force. In addition, job openings are also generated by growth in employment demand. These job openings may be filled by recruitment from many talent pools: new doctorate recipients, experienced doctorates from other labor markets or from outside the labor market (including doctorates from abroad), nondoctorates, etc. In this context, need can be defined as filling future job openings to achieve a particular rate of employment growth or to achieve some alternative goal. The target rate of growth or the alternative goal is a policy decision usually made on normative grounds.

Given this broad context, the committee examines future employment conditions in an effort to estimate need (approximated by job openings) and our ability to meet this need (measured by new Ph.D.'s entering the biomedical sciences workforce). Because job openings can be filled by recruitment from a variety of talent pools, the reader is cautioned that the committee's indicator of our ability to meet this need represents a lower-bound estimate of this ability.

Table 3-3 contains estimates of the future number of job openings to be filled under alternative scenarios about employment growth. Three scenarios are examined: zero growth, 3.6 percent per year (the 1981-1991 compound growth rate for the biomedical science workforce), and 1.8 percent per year (one-half the 1981-1991 compound growth rate). The method used to generate these estimates is a

TABLE 3-3 Committee Estimates of the Average Annual Number of Job Openings Needed to Sustain Various Growth Rates of the Biomedical Work Force[a,b]

Year	Zero Growth Rate Scenario Numbers Needed	Half the Average Growth Rate Scenario[c] Numbers Needed	Average Growth Rate Scenario[d] Numbers Needed
1996-1997	1291	3358	5473
1998-1999	1714	3691	6031
2000-2001	1991	3915	6506

[a]Biomedical work force consists of those employed or on postdoctoral appointments in a biomedical field. Data derived from the NRC Survey of Doctorate Recipients, a sample survey.
[b]Based on multistate life table methods. See Appendix G for methodology.
[c]Half the average referred to in footnote d or 1.8 percent.
[d]Refers to biomedical work force's average annual compound growth rate over the past decade or 3.6 percent (4.25 percent, uncompounded).

variant of demographic cohort-survival models. It generates flows of workers into and out of this work force, and, on the basis of these flows, it generates estimates of changes in the size of this work force.[10] There are, of course, many ways to do multistate life table analysis. The data presented below should be viewed as preliminary work by the committee, which will be explored further by the Panel on Estimation Procedures in the coming months.

Estimates are developed for three time periods: 1996-1997, 1998-1999, and 2000-2001. The estimates are very sensitive to the growth rate assumption, varying from 1,291-1,991 in the zero growth scenario to 5,473-6,506 in the 3.6 percent per year growth scenario. The range is substantially narrower for a given growth rate scenario. The modest increases observed over time for a given rate of growth partially reflect the widely expected increases in deaths and retirements in the late 1990s. Except for the zero growth scenario, they also reflect the growth of the biomedical science work force.[11]

For comparison purposes, Table 3-4 shows the number of new biomedical Ph.D.s entering the biomedical work force through 1990, estimated from the longitudinal SDR.[12] These numbers represent a substantial fraction of the degree production that occurred in these fields, although it does not reflect the employment outcomes of new graduates who may have found employment in other fields or delayed entry into the work force. An estimated 82 percent of the biomedical Ph.D.s entered the biomedical work force

TABLE 3-4 Estimated Number of New Biomedical Science Ph.D.s Entering the Biomedical Science Work Force in Selected Years.

Year	Number[a]
1985-1986	2985
1987-1988	3178
1989-1990	3353

[a]Annual averages.

NOTE: "Biomedical science work force" consists of those employed or on postdoctoral appointments in a biomedical field. The Survey of Doctorate Recipients is a sample survey and subject to sampling error.

SOURCE: NRC, Survey of Doctorate Recipients. (Biennial)

(which includes postdoctorates) during the period 1985-1990.

This level of work force entry, if maintained, could more than meet the need for zero growth, but it will fall considerably short of the number needed to maintain the annual 1981-1991 growth rate. As noted earlier, however, maintenance of this growth rate may be an unrealistic objective. Universities are unlikely to increase faculty size dramatically in the near future, federal spending on biomedical research is not likely to increase in real terms in the near future, and private sector demand (viz., industry) is not likely to increase rapidly in the near future.

The best predictions for economic activity and R&D funding in the near future suggest that demand for basic biomedical scientists will grow slowly at best. Under these circumstances, maintenance of the current rate of entry of Ph.D.s in the biomedical sciences should provide an adequate supply for the years 1996-2001.[13] (See Table 3-3).

The NRSA program supports approximately 5,100 predoctoral students each year in the basic biomedical sciences, although only a fraction complete doctoral degrees in the same year as receiving NRSA support. The number of basic biomedical degree recipients in any year having had NRSA support is unknown but presumed to be small.[14] If current levels of predoctoral NRSA support are maintained and projected, demand for new Ph.D.s is estimated to be 3,400-3,900 per year ("half the average" growth rate scenario, see Table 3-4), then the NRSA program in the basic biomedical sciences will contribute to the preparation of doctoral scientists at a rate which future markets will likely absorb.

Priority Fields

Although market conditions suggest that the demand for basic biomedical scientists may grow more slowly than in the past, we believe that advances in research and continuing requirements to address pressing public health concerns will result in the demand for basic biomedical scientists with quite specific research skills. This does not imply that we need to step up production in all areas; rather, the NRSA mechanism provides an opportunity to increase supply in some areas through relatively small increases in the number of awardees.

There is a continuing need to train young scientists who will have skill and expertise in the well-recognized core biomedical disciplines (e.g., biochemistry, microbiology, and pharmacology) as well as broadly based individuals capable of effective interdisciplinary research. Scientists trained in physical and mathematical sciences and engineering and able to apply knowledge in chemistry, physics, materials science, computational mathematics, and computer science to problems of significance in basic and clinical biomedical sciences will also be required. Much of the current excitement and rapid progress in biomedical science lies at the interfaces between genetics, molecular biology, cellular biology, and developmental biology. The NRSA programs in the basic biomedical sciences appropriately emphasize the kinds of interdisciplinary training required to carry out effective research at these interfaces and to apply new findings to problems in human biology. However, it is also necessary to ensure that there is a cadre of scientists who are knowledgeable in fundamental areas of biomedical science that, for whatever reason, are not at the cutting edge of research at the time. This need, which is perhaps less immediately obvious, is well illustrated by the periodic emergence of new infectious diseases (e.g., Legionnaires disease and cryptosporidiosis) and the recrudescence of diseases, such as tuberculosis, caused by antibiotic- and drug-resistant strains.

ENSURING THE DIVERSITY OF HUMAN RESOURCES

Careers in biomedical research remain attractive to women. At present, between 35 and 45 percent of Ph.D.s awarded in the biomedical sciences have been awarded to women. However, the fraction of women in full-time, independent research positions is still disproportionately low. Moreover, evidence suggests that women rise to the top ranks in academia and industry in fewer numbers than men (NRC, 1991 and 1994). Part of the training process should include explicit mentoring to help women achieve their full career potential.

We need to do much more to increase the number of black and Hispanic students entering research training in the basic biomedical sciences. Continual efforts to attract these students into the NRSA program must be made. Special programs to ensure progress through pre- and postdoctoral training should be encouraged. The Minority Access to Research Careers (MARC) program shows promise as a reliable source of NRSA trainees.

THE NRSA PROGRAM IN THE BASIC BIOMEDICAL SCIENCES

Earlier committees' assessment of the need for basic biomedical scientists and the level of training that should be provided by the federal government under the NRSA programs depended heavily on its analysis of the academic labor market, because that was the dominant sector both in terms of the number of bioscientists employed and the amount of federally sponsored research performed. The number of individuals receiving Ph.D. degrees in the biomedical sciences and the number holding postdoctoral appointments were taken as indicators of supply. Demand indicators were undergraduate and graduate enrollment and the availability of funds for R&D, both of which were perceived to drive the demand for faculty in these fields. Those data, combined with conservative estimates of future trends, were used to make recommendations about the number of trainees needed.

When the committee first convened in 1975, it quickly discerned an oversupply of researchers in the biomedical sciences based on trends in academic employment. Although demand for Ph.D. faculty had experienced rapid growth during the 1960s, by the 1970s the number of students was leveling off, federal funding increases were moderating, and the relatively young faculty members hired during the period of peak expansion were suspected to undergo very little attrition in the near future. In its second report, the committee recommended cutbacks amounting to 30 percent in the number of predoctoral fellows supported annually between fiscal year 1975 and fiscal year 1979, from 6,000 to 4,250, and a level of support of 3,200 postdoctoral fellows. These recommendations were based on evidence of reduced growth in the overall demand for biomedical scientists and continue to affirm the vital role played by the training grant and fellowship programs in training high-quality researchers.

Subsequent reports in 1978, 1979, and 1981 reiterated these recommended levels and suggested that time was needed to evaluate the effects of these cutbacks and further developments in the labor market before new recommendations could be made. By 1981, the committee discerned signs of improvement in the overall job market for biomedical researchers. Academic employment was expanding slightly, largely because of rising enrollments in the biomedical sciences. R&D funding was also beginning to rise, and most promising was the rapid increase of employment in the new biotechnology industry. The committee foresaw continued, strong demand from both small start-up companies and large established corporations that were entering this business. It also saw good employment prospects in the high-priority fields of biostatistics, toxicology, and epidemiology. Nonetheless, it expressed concern about the continued growth of the postdoctoral pool and recommended that the numbers of pre- and postdoctoral awards remain steady at 4,250 and 3,200, respectively.

By 1985 the job market showed clear improvement. For the first time since the reports began, the committee noted a slowing of postdoctoral buildup. The number of Ph.D.s awarded each year also slowed, and although university faculty employment still remained stable, demand in the biotechnology and genetic engineering industries was growing sharply, at more than 9 percent a year. Although the committee did not expect substantial increases in the number of academic positions in the foreseeable future, it did expect retirements to increase markedly by the mid to late 1990s.

In 1989 the study committee projected that an increasing demand for biomedical scientists would exceed the supply through the year 2000. The committee recommended that the level of NRSA predoctoral support be increased to 5,200. The committee also recommended that postdoctoral support be increased gradually as degree production increased.

RECOMMENDATIONS

Total support for the training of basic biomedical scientists increased from just over 9,000 awards in fiscal 1991 to an estimated 9,630 in fiscal 1993 (Table 3-5). This includes about 630 awards for the undergraduate preparation of minority scholars. Predoctoral support is offered primarily through institutional training grants (traineeships), although a limited number of individual fellowships are available. Postdoctoral support in the form of portable fellowships allows eligible applicants to seek advanced preparation in a wide variety of areas. Postdoctoral training grants have emphasized preparation in such areas as genetics, clinical pharmacology, trauma and burn research, and anesthesiology. In making its recommendations in this area, the committee has assumed that the current ratio of predoctoral and postdoctoral support would remain essentially constant, with the majority of support available at the predoctoral level (primarily in the form of traineeships).

Predoctoral Training

On the basis of its review of available information describing current and anticipated market conditions and in consideration of pressing national research needs, the committee strongly endorses the continuation of predoctoral NRSA training programs in the basic biomedical sciences. Although evaluative data remain incomplete, the evidence indicates that these predoctoral training programs remain highly effective in fostering the development and sustaining the health of interdisciplinary graduate programs of benchmark quality, and in catalyzing the entry of highly qualified students into graduate training.

However, the committee is concerned that the current low level of stipend support, $8,800 per year, will erode the impact of these programs and their ability to attract the most talented students. We recommend that stipends be increased incrementally over a 2-3 year period to $12,000 for all predoctoral awardees. We consider the recommended increase in stipend to be of higher priority than any possible increases in number of trainee slots, and therefore recommend that the number of predoctoral awards remain at FY 1993 levels during this period. The committee recognizes that these recommendations are being made in an era of

TABLE 3-5 Aggregated Numbers of NRSA Supported Trainees and Fellows in Basic Biomedical Sciences for FY 1991 through FY 1993

Fiscal Year	Level of Training	TOTAL	Type of Support	
			Traineeship	Fellowship
1991	Number of awards	9,021	7,199	1,822
	Predoctoral	4,593	4,313	280
	Postdoctoral	3,861	2,319	1,542
	MARC Undergraduate	567	567	-
1992	Number of awards	9,317	7,477	1,840
	Predoctoral	4,777	4,487	290
	Postdoctoral	3,910	2,360	1,550
	MARC Undergraduate	630	630	-
1993	Number of awards	9,633	7,740	1,893
	Predoctoral	5,171	4,811	360[a]
	Postdoctoral	3,836	2,303	1,533
	MARC Undergraduate	626	626	-

a Includes minority scholars supported through the National Minority Fellowship Program. See Appendix E.

NOTE: Based on estimates provided by the National Institutes of Health. See Summary Table 1.

fiscal constraint. Should additional funds become available for research training in the basic biomedical sciences, the NIH might wish to consider expanding NRSA support in this area.

RECOMMENDATION: The committee recommends that the number of predoctoral trainees and fellows supported annually in the basic biomedical sciences be maintained at 1993 levels or approximately 5,175 each year (Table 3-6).

Postdoctoral Training

Postdoctoral research training sharpens the technical and intellectual skills of the doctoral-level scientist and provides important (and frequently used) opportunities for cross-disciplinary training as preparation for undertaking a career as an independent investigator.

The committee is concerned, however, that persistent low-level stipends may discourage qualified applicants from seeking postdoctoral training through NRSA support. Thus, to permit NIH to introduce further and more realistic changes in stipend levels at the postdoctoral level, the committee recommends that the number of postdoctoral awards be maintained at fiscal 1993 levels (Table 3-6). Should, however, additional program funds become available for postdoctoral training in the basic biomedical sciences, the NIH may also wish to expand support for postdoctoral training.

RECOMMENDATION: The committee recommends that the number of postdoctoral trainees and fellows supported annually in the basic biomedical sciences be maintained at 1993 levels or 3,835 each year.

Minority Access to Research Careers

Current federal efforts to attract minority group members to careers in the basic biomedical sciences include undergraduate support through the MARC program. The core of this program is the Honors Undergraduate Program launched in fiscal 1977 to support college juniors and seniors. In fiscal 1993 approximately 630 individuals received undergraduate support (see Table 3-5).

As noted in Chapter 9 of this report, NIH recently initiated an 18-month study of the career outcomes of the MARC program. The committee endorses continuation of funding for this program at current levels to support the training of individuals in the basic biomedical sciences until the NIH assessment is complete and information is made available to subsequent NRC study committees.

RECOMMENDATION: The committee recommends that the number of NRSA awards made available through the MARC undergraduate program for research training in the basic biomedical sciences be maintained at about 630 awards each year.

TABLE 3-6 Committee Recommendations for Relative Distribution of Predoctoral and Postdoctoral Traineeship and Fellowship Awards for Basic Biomedical Sciences for FY 1994 through FY 1999

Fiscal Year	Level of Training	TOTAL	Type of Support	
			Traineeship	Fellowship
1994	Recommended number of awards	9,640	7,745	1,895
	Predoctoral	5,175	4,815	360
	Postdoctoral	3,835	2,300	1,535
	MARC Undergraduate	630	630	-
1995	Recommended number of awards	9,640	7,745	1,895
	Predoctoral	5,175	4,815	360
	Postdoctoral	3,835	2,300	1,535
	MARC Undergraduate	630	630	-
1996	Recommended number of awards	9,640	7,745	1,895
	Predoctoral	5,175	4,815	360
	Postdoctoral	3,835	2,300	1,535
	MARC Undergraduate	630	630	-
1997	Recommended number of awards	9,640	7,745	1,895
	Predoctoral	5,175	4,815	360
	Postdoctoral	3,835	2,300	1,535
	MARC Undergraduate	630	630	-
1998	Recommended number of awards	9,640	7,745	1,895
	Predoctoral	5,175	4,815	360
	Postdoctoral	3,835	2,300	1,535
	MARC Undergraduate	630	630	-
1999	Recommended number of awards	9,640	7,745	1,895
	Predoctoral	5,175	4,815	360
	Postdoctoral	3,835	2,300	1,535
	MARC Undergraduate	630	630	-

NOTES

1. The slowdown in the rate of growth between 1989 and 1991 was accompanied by an absolute decline in academia. In part, this may reflect methodological changes that occurred in the Survey at that time. However, it may also reflect a weakening in demand, particularly in the academic sector.

2. Includes both citizens and noncitizens, where citizens includes both native-born and naturalized citizens.

3. Special run, Survey of Doctoral Recipients (SDR). SDR is a biennial survey of a sample of scientists and engineers conducted by the NRC behalf of the federal government.

4. Special run, SDR. A number of authors have also observed that women are generally underrepresented in tenure-track faculty positions relative to the numbers among doctoral recipients (NRC, 1981; Zuckerman et al., 1992).

5. The Doctorate Records File is a compilation of responses to the Survey of Earned Doctorates, which has been conducted each year since 1958 by the NRC's Office of Scientific and Engineering Personnel and its predecessor organizations. Questionnaires, distributed with the cooperation of the graduate deans of U.S. universities, are filled in by graduates as they complete requirements for their doctoral degrees. The doctorates are reported by academic year and include research and applied-research doctorates in all fields. See Ries and Thurgood, 1993.

6. Much more work is needed to document fully the impact of foreign participation on the U.S. science and technology work force. Owing to the nature of many of our data sources, we are unable to determine at this time how many non-U.S. citizens who earn doctoral degrees in this country remain in the U.S. and we know very little about the careers and research contributions of non-U.S. citizens to the U.S. research effort regardless of the origin of their doctoral degrees.

7. The Panel on Estimation Procedures considered numerous "short run" indicators such as wage adjustments for young workers relative to older workers, relative tenure-earning ratios, and job openings. However, owing to limitations of time and resources, the Panel and Committee restricted these analyses to more readily available information.

8. The postdoctoral appointment has become an essential component of advanced training in most subfields of the basic biomedical sciences. Past studies by the National Research Council (Garrison and Brown, 1984, for example) have suggested that individuals with postdoctoral training enter more productive research careers than those individuals without postdoctoral training. See Appendix A of this report for a brief summary of the career outcomes studies of NRSA postdoctoral appointees. Nonetheless, the expansion of postdoctoral appointments in the basic biomedical sciences has been identified by some researchers as an indicator of job shortages in some component fields (Coggeshall, et al., 1978; NRC, 1981).

9. An alternative interpretation of this finding is that the relative supply of biomedical scientists increased faster than that of other scientists and engineers. The decline in starting wages would thus result from an increase in relative demand.

10. The flows are generated from multistate life tables. These tables are based on matrices of age-specific transition rates estimated from the Survey of Doctoral Recipients historical data. These rates are assumed to remain constant over time. For a more detailed description of the methodology, see Appendix G. This analysis will be reviewed closely by the Panel on Estimation Procedures along with other approaches to the estimation of national needs relative to human resource training and policies.

11. Recall that in developing these estimates, it is assumed that age-specific separation rates remain stable. There is, however, evidence of a strong positive relationship between these rates and age (NRC, 1989). Given this relationship, the upward trend in the numbers may also be reflecting the expected aging of this population.

12. SDR. See note 3.

13. The estimate is presented as a minimum value because these job openings could also be filled by recruiting workers with degrees and training in closely related fields or workers from abroad.

14. On the basis of information gathered from the National Science Foundation the committee estimates that less than 15 percent of graduate students in the life sciences received NRSA support in FY 1990.

REFERENCES

Coggeshall, P., J. C. Norvell, L. Bogorad, and R. M. Bock
    1978   Changing postdoctoral career patterns for biomedical scientists. *Science* 202:487-493.

Matyas, M. and L. S. Dix (eds)
    1992   *Science and Engineering Programs: On Target for Women?* Washington, D.C.: National Academy Press.

National Science Foundation (NSF)
    1990   *Immigration of Scientists and Engineers to the United States: A Literature Review.* Science Resources Studies Division. Mimeographed. March. Washington, D.C.

National Research Council (NRC)
    1994   *The Funding of Young Investigators in the Biological and Biomedical Sciences.* Washington, D.C.: National Academy Press.
    1981   *Postdoctoral Appointments and Disappointments.* Washington, D.C.: National Academy Press.
    1989   *Biomedical and Behavioral Research Scientists: Their Training and Supply, Volume I: Findings.* Washington, D.C.: National Academy Press.
    1991   *Women in Science and Engineering. Increasing Their Numbers in the 1990s.* Washington, D.C.: National Academy Press.
    1994   *Women Scientists and Engineers Employed in Industry: Why So Few?* Washington, D.C.: National Academy Press.

Office of the President of the United States
    1993   *Economic Report of the President.* Washington, D.C.: U.S. Government Printing Office.

Ries, P. and D. H. Thurgood
    1993   *Summary Report 1992: Doctorate Recipients from United States Universities.* Washington, D.C.: National Academy Press.

Zuckerman, H., J.R. Cole, and J.T. Bruer
    1991   *The Outer Circle: Women in the Scientific Community.* New York: W. W. Norton and Company.

CHAPTER FOUR

BEHAVIORAL SCIENCES PERSONNEL

The United States leads the world in behavioral research, and there is now increasing recognition, especially among policymakers, that the solution to many of the worst problems facing the country are primarily behavioral in character. The behavioral sciences[1] yield knowledge to deal with many of these problems.

Behavior ranges from individual skills (for example, the lack of self-control implicated in violent behavior) to cognitive mediators of risky behavior (e.g., an individual's perceptions of invulnerability relative to risk-taking behaviors that affect health) to community-level phenomena (e.g., an individual's belief in the legitimacy of medical prescriptions derived from beliefs shared with a group). Every area of major health risk in this country can be informed by behavioral science research. Moreover, in some areas of great risk—such as child abuse, substance abuse and crime, and acquired immune deficiency syndrome (AIDS)—large scale behavioral programs are being developed, and advances in demographic and population studies have enhanced our understanding of "longevity".

Since issuing its first report on the National Research Service Awards (NRSA) program in 1975, the National Research Council (NRC) has strongly endorsed the continuation of support for research training in the behavioral sciences through the NRSA program. This occurred in recognition of the inclusion of the Alcohol, Drug Abuse, and Mental Health Administration (ADAMHA) in the restructured research training authority (P.L. 93-348). The behavioral sciences also received attention because of their critical role in exploring the underlying processes of normal human development and aging; for their role in outlining the causes, etiology, and treatment of diseases involving neurological and sensory processing; and for their role in prevention research. In October 1992 the research components of the three former institutes of ADAMHA joined the National Institutes of Health (NIH) signaling an historic shift (Goodwin, 1993) in the nation's view of the subtle interactions between changes in behavior and health and dysfunction.

Recommended levels of support in the behavioral sciences have been predicated on the belief that NRSA awards leverage the production of small numbers of highly skilled workers whose research is conducted in the national interest. Indeed, follow-up studies have found most former NRSA awardees actively engaged in research (see Appendix A).

The number of degree recipients having had NRSA support remains low and, therefore, does not drive the supply of behavioral science doctorates. Nonetheless, the committee this year has paid particular attention to the market for behavioral scientists for reasons related to the health of the overall behavioral research enterprise. As noted in previous chapters (2 and 3), the committee has replaced earlier committees' analyses of supply-and-demand models with the use of new techniques which involve multistate life table analysis of future supply and a separate assessment of short-term indicators of current and past market conditions.

The committee was gratified to learn, for example, that despite severe funding cuts in the early 1980s, behavioral scientists continue to find productive employment although overall Ph.D. production in these areas has slowed somewhat in recent years. Based on this information, and information gathered from a variety of other sources, we conclude that the demand for behavioral science personnel will grow slowly as a whole but that the few behavioral scientists receiving NRSA support can expect to enter into productive employment upon completion of their doctoral or postdoctoral studies.

While these analyses provide an important backdrop against which the committee's judgment about future needs for behavioral research personnel can be assessed, they represent but one of the several dimensions of "needs." Cognizant of the important advances in behavioral research that promise to enhance the health and well-being of all citizens and confident that NRSA awardees will have the opportunity to contribute to the national health effort, the committee has concluded that the present modest program of NRSA support in the behavioral sciences should essentially double in the next few years. However, because of our concern with current low levels of stipend support, we have adjusted our goals to permit immediate expansion of training stipends throughout the NRSA program. We believe, however, that the national call for skilled behavioral science investigators should be answered swiftly and strongly through an increase in training support in this area.

ADVANCES IN RESEARCH IN THE BEHAVIORAL SCIENCES

The knowledge base in the behavioral sciences has reached the point where effective utilization of its findings by physicians, clinical psychologists, nurses, and social workers can have a very significant impact on health-related problems in our society. Advances in basic research in behavioral science dovetail well with national needs.

This is a most exciting and intellectually stimulating time in behavioral science. Work on both basic and applied problems—and work done at many levels of analysis, from brain-behavior relations to the study of disease processes in human populations—is making great progress. We cite three examples in more detail to illustrate the intellectual excitement in the field.

Health and Behavior

Much of the social and behavioral research supported by the National Institutes of Health represents research on "health and behavior". Current estimates place NIH support for research in this area at about 8 percent of total NIH R&D support (COSSA, 1994). The "NIH Implementation Plan for Health and Behavior Research" (NIH, 1993) outlines, furthermore, what the Institutes and Centers consider to be optimal spending levels in this area during the next 5 to 10 years.

Expanded support for research on health and behavior reflects an increasing recognition by the health research community that social and psychological factors play a significant role in the natural history of disease, prevention of disability and illness, and promotion of recovery. As the Director of NIH stated in a 1991 report on the same subject (NIH, 1991):

> Our research is teaching us that many common diseases can be prevented, and others can be postponed or well-controlled, simply by making positive life style changes. For these reasons, intensifying such research and encouraging all Americans to make health-enhancing behaviors a part of their daily lives has taken on more and more importance in our efforts to conquer disease.

The domain of research on health and behavior is broad and the enormity of its knowledge base is daunting, but as Adler and Matthews (1994) suggest, many of the concepts in recent years pertain to three essential questions:

1. Who becomes sick and why?
2. Among the sick, who recovers and why?
3. How can illness be prevented or recovery promoted?

To facilitate the answers to these and other related questions, the NIH Reauthorization Act of 1993 established the Office of Behavioral and Social Sciences Research and called for a report to the U.S. Congress on the extent to which the Institutes of Health conduct and support research in the component disciplines. Because the Office is still being organized within the National Institutes of Health, Howard Silver and the staff of the Consortium of Social Science Associations (COSSA) recently prepared a detailed, although preliminary, summary illustrating NIH funding priorities in the area of health and behavior. For example, the COSSA summary notes that the National Heart, Lung, and Blood Institute (NHLBI) supports a variety of activities—primarily through the Behavioral Medicine Branch—on disease prevention, etiology, diagnosis and treatment of cardiovascular diseases. The National Institute of Allergy and Infectious Diseases (NIAID), which is concerned increasingly with the transmission of AIDS, supports only a modest amount of social and behavioral research but has expressed interest in psychosocial factors affecting medical treatment of compliance. The National Institute of Environmental Health Sciences, to give another example, supports research on the effects of environmental agents on human health and well-being with particular attention to behavioral and neurological effects of exposure to toxic substances.

Variables that have been explored by research scientists over the years can be categorized in any number of ways, but include research on factors that arise from the social environment which contribute to disease. Such factors include: stress (Cox and Gonder-Frederick, 1992; Beardsley and Goldstein, 1993) or social isolation/social connectedness (Cohen, 1988; Reynolds and Kaplan, 1990). Individual dispositional factors are also thought to contribute to disease onset and recovery, such as: hostility/Type A personality (Siegrist et al, 1990; Matthews et al, 1992); depression/exhaustion (Hahn and Pettiti, 1988; Markovitz et al, 1991); neuroticism and negative affect (Costa and McCrae, 1987; Salovey and Birnbaum, 1989); and optimism/self-esteem (Schreier and Carver, 1992; Brown and McGill, 1989).

Advances in research on health and behavior often emerge from multidisciplinary studies involving social, behavioral and biomedical scientists including clinicians. As federal funding priorities increasingly emphasize research in this area, we can anticipate that new research training opportunities will emerge, perhaps requiring consideration of new research training arrangements which emphasize cross-disciplinary connections.

Learning and Memory

Memory is dissociated into processes or systems that are fundamentally different. For example, amnesic patients with brain injury or disease exhibit severe inabilities to recall and recognize recent events and have difficulty learning new facts. However, these patients possess some relatively intact learning and memory on tasks such as manual-dexterity learning, which they perform as well as do healthy and uninjured people, even though they may have no conscious memory of having performed the task before. This evidence—that some kinds of learning can proceed normally even when the brain structures that mediate conscious remembering are damaged—supports the general proposition that there are distinct, dissociated types of memory.

Memory is an active process of seeking and reconstructing information, not a passive recording and reproducing of events. Thus, expectations of what things should look like or the way events should happen influence what people notice and remember. For example, after listening to a story presented in jumbled order, people still tend to remember it as being told in proper sequence, following certain widely accepted scenarios for what constitutes a story. People also tend to pay little attention to the details of routine situations. Consequently, people often remember that the most probable things happened even when they did not. This phenomenon has been demonstrated in the context of eyewitness court testimony. Memory of an event can be modified or distorted by how questions about the event are posed. Such experimental findings have important theoretical implications for understanding the formal structure of memory and they have practical implications for legal proceedings.

Recent progress in the broad field concerned with brain substrates of learning and memory (behavioral and cognitive neuroscience) has been impressive. The essential circuits and probable loci of memory traces have been largely identified and mechanisms are being characterized for several invertebrates (e.g., Clark and Schuman 1992; Kandel, 1976); for learned fear in mammals, including both behavioral and autonomic measures (amygdala) (Hitchcock and Davis, 1986; McGaugh, 1989); and for classical conditioning of discrete behavioral responses (skilled movements) in mammals, including humans (cerebellum) (Daum et al., 1993; Thompson, 1986). Progress is rapid in determining the role of the hippocampus and associated cortical areas in contextual-spatial and declarative memory (Meuner et al., 1993; O'Keefe and Nadel, 1978; Squire, 1992).

A particularly important application of this basic research is in aging and Alzheimer's disease. The procedural memory system concerned with motor skill learning declines with normal aging, with concomitant loss of neurons in the cerebellum. Declarative or experiential memory shows only a modest decline in normal aging but a dramatic and profound impairment in Alzheimer's disease. The brain substrates of this declarative memory system (hippocampus and associated cortical areas) are particularly vulnerable and show progressive and profound deterioration in the disease. Thanks to the well-developed fields of learning and memory and abilities testing, we now have very sensitive diagnostic tools to identify the earliest stages of putative Alzheimer's and other dementias.

Early childhood is a period rich in experience in which young children are capable of acquiring, registering, and recalling events and episodes as well as recognizing places and people. They can also vividly react to impressions, manifest pain and pleasure, and express love, jealousy, and other passions. It is thus astonishing that this period is usually lost to memory entirely except for a few fragments. Recent work in developmental-cognitive psychology and cognitive neuroscience suggests that infantile amnesia is not due to repression, as Freud hypothesized but rather occurs because of the length of time necessary to develop the requisite cognitive skills and the concomitant developmental schedule of the brain substrate of declarative (experiential) memory. The hippocampal and cortical memory system is not fully mature until about the age of 3 (Bachevalier, 1991). These data raise serious questions concerning the current rash of therapy- and hypnosis-related instances of patient accusations of infantile assault, based on presumed recovery of infantile memories (for example, *Time* Magazine, November 1993). Over and above the issue of suggestibility is the well-documented evidence that the original memory of an experience can be modified substantially long after the fact by how questions about the experience are phrased (Loftus and Loftus, 1976).

Signal Detection and Medical Imaging

Signal-detection theory is an outgrowth of behavioral science, first noted in the field of hearing. A fundamental problem in hearing is how to detect the presence of a signal in noise. Two extreme listener strategies are to respond positively with great uncertainty, leading to many false positive responses, and to respond positively only with great certainty, leading to many false-negative responses. In a brilliant mathematical and behavioral analysis, Swets,

Green, and many other behavioral scientists developed a comprehensive theory of signal detection that makes it possible to establish probability levels of these two types of errors and to manipulate them by establishment of criterion levels of judgment (see, Green and Swets, 1966).

The practical applications growing from this basic research have gone beyond sensory problems. Modern electronic and computer technologies for image enhancement have wide applications today, ranging from enhancement of space-probe images to brain imaging to computer-reading of X-ray films. Computational techniques combined with various medical imaging procedures are making it possible to see specific organs and internal body parts volumetrically and in ever-increasing detail. Computer-based imaging techniques include computed axial tomography, nuclear magnetic resonance imaging, positron emission tomography, and digital subtraction angiography. Some techniques, by assigning different degrees of transparency to different layers of bones and tissue, let the observer not only see them but look through them and see what lies within or behind them (Fitzgerald, 1989).

Signal-detection theory has addressed the diagnostic power and cost effectiveness of these imaging techniques, relative to more conventional and less expensive X-ray technology (Swets, 1979 and 1988; Swets and Pickett, 1982; Metz, 1986). As medical imaging technology continues to develop, the need for studies evaluating the effectiveness of specific instruments and procedures will increase. This is so especially in view of the considerable cost of the equipment involved and because diagnostic images produced will not always be interpreted by image interpreting specialists but by physicians who view certain types of images only occasionally. There is a need also for the development and evaluation of procedures (including computer-aided procedures) to help clinicians, especially those with limited experience in image interpretation, to extract from medical images the diagnostically relevant information they contain (Getty et al., 1988; Swets et al., 1991).

Another example of current practical application is the cochlear implant. This application, like many from behavioral science, is interdisciplinary in that it involves neuroscience, otology, physics, engineering, and behavioral science, specifically the psychophysics of hearing and methods of training.

Signal-detection theory is one aspect of the broader field of decision theory, a set of well-developed methods to evaluate decision-making in situations where uncertainty is inherent. These methods have been applied, with considerable success, in a variety of medically related situations and have proved to be extremely helpful in improving training procedures that both accelerate training and increase the overall quality of medical diagnosis (Luce et al., 1989). A dramatic illustration of such an effect is given by McNeil et al. (1982), who asked people to imagine that they had lung cancer and had to choose between two therapies: surgery or radiation. Each therapy was described in some detail. Then, some subjects were presented with the cumulative probabilities of surviving for various lengths of time after each type of treatment. Other subjects received the same cumulative probabilities expressed in terms of dying rather than surviving (e.g., instead of being told that 68 percent of those having surgery will have survived after 1 year, they were told that 32 percent will have died). Framing the statistics in terms of dying dropped the overall percentage of subjects choosing radiation therapy over surgery from 44 percent to 18 percent. Of the three groups of subjects used in the study—patients, students, and physicians—this effect was found to be the strongest among physicians.

ASSESSMENT OF THE CURRENT MARKET FOR BEHAVIORAL SCIENTISTS

The labor market for researchers in the behavioral sciences was robust in the 1980s. The behavioral sciences work force increased by 35 percent between 1981 and 1991, climbing from approximately 49,000 to nearly 67,000 (Figure 4-1).[2] However, over half of this growth was accounted for by clinical psychologists, who are less involved in research and development (R&D) than are their nonclinical counterparts (Pion, 1993).[3] Although psychologists have been fully employed and utilizing their research skills in recent years, there has been a shift in employment prospects from nonclinical to clinical fields.

Accompanying this growth were notable changes in the composition and employment distribution of these workers. Among these changes were the growing prominence of women and members of racial and ethnic minority groups and the increasing importance of the nonacademic sectors in providing employment opportunities. The latter change

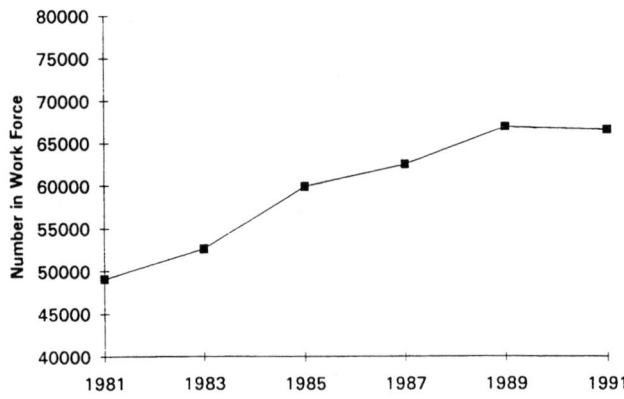

FIGURE 4-1 U.S. behavioral science employment, 1981-1991. See Appendix Table F-13.

reflects in part the rapid employment growth of clinical psychologists, who are primarily employed in nonacademic positions.

Women have been becoming an increasingly important component of the labor market for behavioral scientists. Nearly 40 percent of this work force was female in 1991 compared with 27 percent in 1981 (Figure 4-2). The females tend to be concentrated in the younger age groups. About two-fifths of the female behavioral scientists were younger than 40 in 1991 compared with, at most, one-fourth of the comparable males.[4]

The U.S. work force has become more racially diverse over the years and the behavioral science work force also reflects these changes. In 1991 nearly 8 percent of employed behavioral science Ph.D.s represented individuals from a racial minority group (Table 4-1). In 1979 minorities represented about 4 percent of these Ph.D.s. Most of the growth occurred for blacks. The progress in ethnic diversity is less dramatic. In 1991 Hispanics represented about 2 percent of the behavioral scientists. In 1979 the comparable statistic was roughly 1 percent.

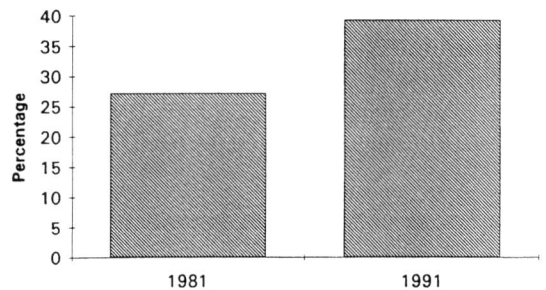

FIGURE 4-2 Fraction of the U.S. behavioral science work force who are women, 1981 and 1991. See Appendix Table F-13.

The behavioral work force is aging. The median age of behavioral scientists has increased from almost 40 in 1981

TABLE 4-1 Racial/Ethnic Composition of Employed Behavioral Science Ph.D.s: 1981 and 1991

	1981[a]		1991[b]	
	Number	Percent	Number	Percent
Race				
TOTAL	55,821	100.0	88,340	100.0
White	53,506	95.9	83,533	94.6
Black	1,102	2.0	2,590	2.9
Asian/Pacific Islander	1,013	1.8	1,924	2.2
Other (Incl. Native American)	200	0.4	293	0.3
Ethnicity				
TOTAL	54,742	100.00	87,686	100.0
Hispanic	859	1.6	1,931	2.2
Non-Hispanic	53,883	98.4	85,755	97.8

[a]For those who responded in 1981. Race nonresponse was 222 in 1981 and ethnic nonresponse was 1,301.
[b]For those who responded in 1991. Race nonresponse was 229 in 1991 and ethnic nonresponse was 883.

NOTE: Employed behavioral science Ph.D.s are those with a behavioral science Ph.D. in behavioral science fields, regardless of employment field. Estimates are subject to sampling error. Comparisons between 1991 estimates and those of earlier years should be made with caution due to changes in survey methodology. Prior to 1991, the SDR collected data by mail methods only. In 1991, the survey had both a mail component and a telephone follow-up component. In this table, 1991 estimtes are based on "mail-only" data to maintain greater comparability with earlier years. Totals may not add up to 100 due to rounding.

SOURCE: NRC, Survey of Doctorate Recipients. (Biennial)

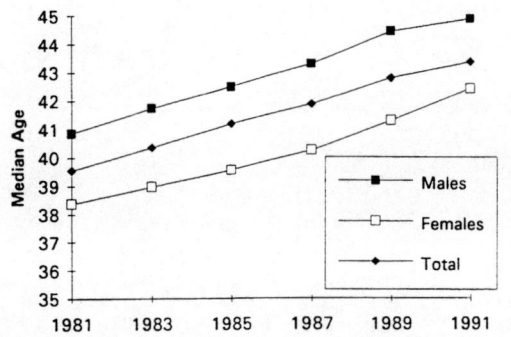

FIGURE 4-3 Median age of the U.S. behavioral science work force by gender, 1981-1991. See Appendix Table F-13.

to a little over 43 in 1991 (Figure 4-3). In part, this reflects trends in degree production. The number of degrees awarded in the behavioral sciences actually fell for a period of time and showed no strong trend in the latter part of the decade.

Over 95 percent of employed behavioral science Ph.D.s were U.S. citizens in 1991, unchanged from comparable numbers for 1981 (Figure 4-4). Almost all U.S. citizens in the behavioral sciences were native born, rather than naturalized (94 percent). On the basis of these data, we can conclude that immigration is not as important a source of supply to the behavioral sciences as it is to other fields, particularly the mathematical and physical sciences and engineering. The comparable 1991 numbers for all science and engineering fields were 93 percent for U.S. citizens and 83 percent for native-born citizens.[5]

Postdoctoral appointments are less important in the labor market for behavioral scientists than for biomedical scientists. Fewer than 1 percent of the behavioral science work force were employed in postdoctoral positions (in contrast to 7 percent in the biomedical sciences). Proportionately more women than men held postdoctoral appointments (1 percent vs. 0.4 percent).[6]

The academic sector has not been as important as a source of employment for behavioral scientists in recent years as it has been for most other science fields. And it is becoming even less important. Most behavioral scientists were employed in settings outside the academic sector, with strong representation in industry and in hospitals and clinics.[7] As noted earlier, this reflects in part the inclusion of clinical psychologists in these statistics. Figure 4-5 details the trends. In 1981 roughly 55 percent of the behavioral scientists were employed in the academic sector. By 1991

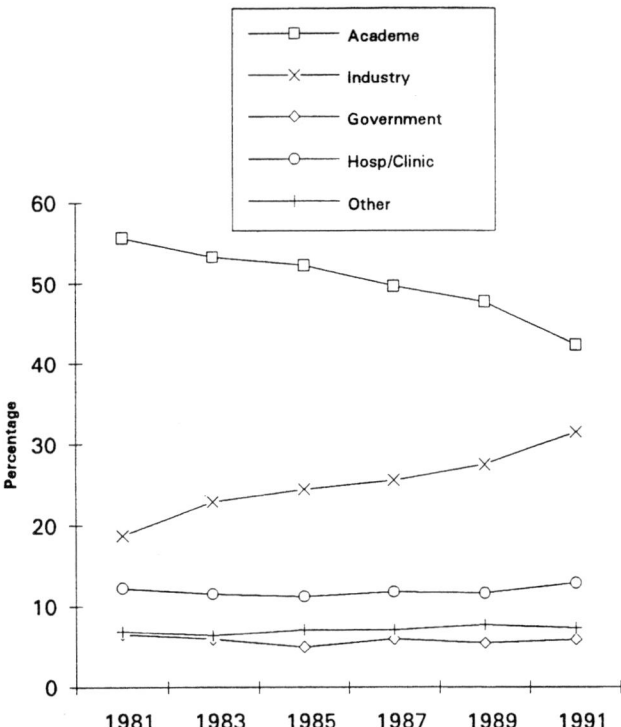

FIGURE 4-5 Employment sector of the U.S. behavioral science work force, 1981-1991. See Appendix Table F-15.

this percentage had fallen to 42 percent. The shift from the academic sector was almost entirely absorbed by the industrial and clinical sectors, which rose from 19 percent in 1981 to 32 percent in 1991.

Degree Production and Career Patterns

There are many sources of talent available to the market for behavioral sciences, and they operate in rather complex ways. The distinctive role of clinical psychologists further complicates interpretation of labor market data. The major source of new behavioral science talent has traditionally been our nation's university system, but some jobs in this

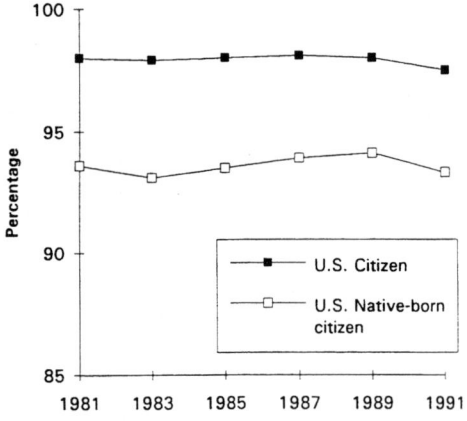

FIGURE 4-4 Citizenship status of employed behavioral science Ph.D.s, 1981-1991. See Appendix Table F-14.

market are filled by workers with degrees in closely related fields or in special Ph.D. programs of research in social work, nursing, and health services. Unlike the biomedical labor market, however, immigration does not appear to be a major supplier of talent.

Degree Production

The most readily available source of information describes degree production from U.S. universities, and the committee summarizes this information below. In contrast to the trends observed for the biomedical sciences, the data reveal a significant downward trend in degree production. The annual number of degrees produced in the behavioral sciences fell between 1981 and 1989, rose between 1989 and 1991, and declined slightly between 1991 and 1992. The net result was that degree production declined by about 12 percent (from 4,149 to 3,647) between 1981 and 1992 (Figure 4-6). This downward trend in degree production starkly contrasts with the increases experienced during this period by the biomedical sciences and by all fields of science and engineering.[8]

Significant progress has been made in achieving gender diversity, and the behavioral sciences have been in the forefront of this movement. Women have represented more than half of this degree production since 1984 (Figure 4-7). The share of degrees granted to women increased from 44.2 percent to 58.7 percent between 1981 and 1992. Although total degree production in the behavioral sciences remained essentially unchanged during this period, the number of degrees granted to women rose 17 percent, from roughly 1,832 to about 2,142.

Less progress has been made with respect to race and ethnic diversity, however. Whites constituted about 90 percent of degree production in 1981-1992 (Table 4-2). There were small increases in the shares awarded to Asians and Hispanics. The black share remained virtually unchanged.

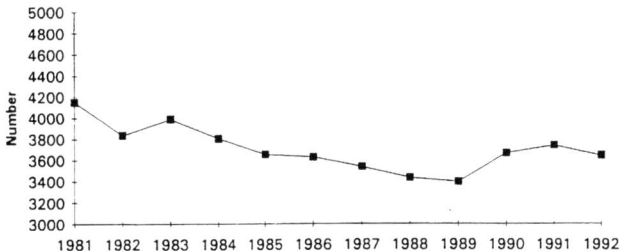

FIGURE 4-6 Behavioral science Ph.D. production, 1981-1992. NOTE: Data limited to U.S. citizens and permanent residents. See Appendix Table F-16.

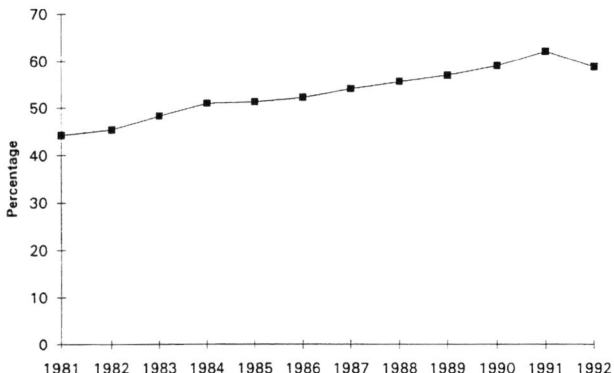

FIGURE 4-7 Fraction of behavioral science Ph.D. degrees earned each year by women, 1981-1992. NOTE: Data limited to U.S. citizens and permanent residents. See Appendix Table F-16.

TABLE 4-2 Behavioral Ph.D. Production Over Time, by Race and Ethnicity

		1981	1982	1983	1984	1985	1986	1987	1988	1989	1990	1991	1992
Total	N	4022	3765	3896	3743	3586	3572	3477	3383	3353	3610	3681	3594
White	%	92.0	91.3	91.5	90.9	90.9	90.6	90.2	90.1	89.7	89.2	88.5	89.1
Black		3.9	4.2	3.8	4.3	4.1	4.0	3.4	3.9	4.1	4.0	4.6	3.9
Hispanic		2.3	2.6	2.9	2.9	2.9	3.3	3.5	3.6	3.7	1.4	4.2	4.0
Asian		1.6	1.4	1.5	1.6	1.8	1.7	2.3	2.0	2.2	2.1	2.3	2.4
Native American		0.3	0.5	0.3	0.2	0.4	0.4	0.5	0.4	0.4	0.6	0.4	0.6

NOTE: Cases with missing data are excluded. Data limited to U.S. citizens and permanent residents.

SOURCE: NRC, Survey of Earned Doctorates. (Annual)

Unlike biomedical degree recipients, very few new doctorates are not U.S. citizens and there is no strong trend suggesting that this situation is changing. Roughly 94 percent of all new doctorates in the behavioral sciences were U.S. citizens in 1981. This figure had fallen to approximately 89 percent in 1992. As noted earlier, these statistics do not imply any large changes in the future composition of the behavioral science work force with respect to citizenship status (Figure 4-8).

Career Patterns

Given the objective of the NRSA awards—to produce research scientists—it is useful to have some notion of the number of years over the course of a career that these scientists remain engaged in R&D. The effectiveness of the program will vary with this number. As noted in Chapter 3, the Survey of Doctoral Recipients—a longitudinal survey that tracks doctorates in the sciences, engineering and humanities biennially—provides useful information on employment patterns, including postdoctoral work. This survey has the potential for illuminating career patterns of behavioral scientists. Thus, the Panel on Estimation Procedures will examine more closely the feasibility of estimating such patterns.

Market Conditions

This section presents selected indicators of short-term market conditions, which include unemployment and underemployment rates, postgraduation commitments of new doctorates, and relative salaries.[9]

Unemployment and Underemployment

As noted in Chapter 3, the most commonly used short-term indicator of labor market conditions is the unemployment rate. It is not as meaningful as an indicator of these conditions for highly skilled workers, however. Because such workers are usually able to find jobs—even in times of weak demand—the issue is not whether the worker has a job, but whether the job is fully utilizing his or her skills. For this reason, the committee has also compiled information on underemployment: i.e., workers who are working part time but would prefer full-time jobs and workers who have jobs that are outside of science and engineering and who indicate they took these jobs because they could not find work in science and engineering.

Figure 4-9 summarizes the unemployment rates. Given the recent publicity about the weak state of demand in the physical sciences, comparable rates for physical scientists are also included so that the reader can assess conditions in behavioral science labor markets relative to those in the physical sciences.

Rates of unemployment have generally been in the range of 1 percent, with little variability. The rate of underemployment is also relatively low (Figure 4-10).

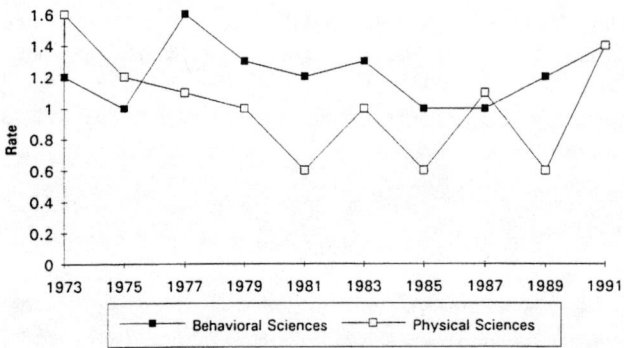

FIGURE 4-9 Unemployment rates for behavioral and physical sciences Ph.D.s, 1973-1991. See Appendix Table F-18.

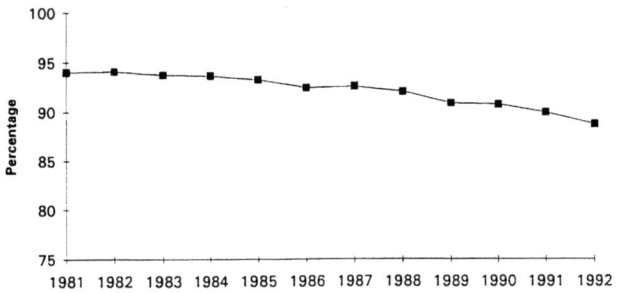

FIGURE 4-8 Fraction of behavioral science Ph.D. degrees earned each year by U.S. citizens, 1981-1992. See Appendix Table F-17.

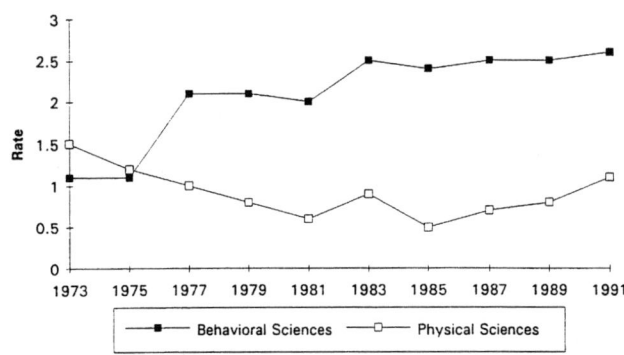

FIGURE 4-10 Underemployment rates for behavioral and physical sciences Ph.D.s, 1973-1991. See Appendix Table F-19.

Postgraduation Commitments

Postgraduation plans of new doctorates may also reflect market conditions. In particular, the percentage of new Ph.D.s who indicate that they have definite commitments at the time they are completing their requirements for the degree can reflect the strength of demand. When demand is weak, this percentage will fall; when demand is strong, this percentage will rise.

Figure 4-11 summarizes these plans for 1975-1992. To provide a comparative base, similar information is provided for degree recipients in the physical sciences, a field thought to be suffering currently from weak demand. The data reveal a declining trend in the percentage of new behavioral science graduates who report definite employment or postdoctorate commitments when they receive their degrees. In contrast the physical sciences display an upward trend for 1975-1981 and a slow decline beginning in 1981

positions (excluding postdoctoral positions) relative to comparable salaries of all employed science and engineering doctorates, age 30-34, are presented in Figure 4-12.[10] Since 1983 these salaries have been declining relative to those of comparable scientists and engineers in all fields. This supports the hypothesis that the relative demand for behavioral scientists has been declining. However, the behavioral sciences represent a heterogeneous collection of disciplines with quite diverse employment paths. Clinical psychologists, for example, have increased their numbers and typically pursue career paths in sectors other than academia.[11] Also, social scientists have for many years found research opportunities in business and industry, including self-employment. As a result of these differences in employment opportunities, starting salaries are likely to be influenced by "compositional" effects. Thus, these data must be interpreted with some caution.

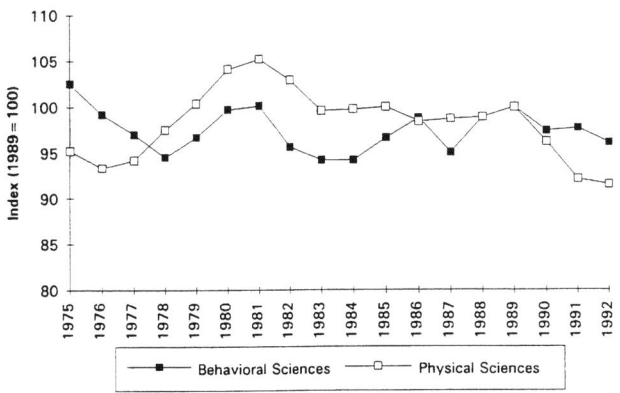

FIGURE 4-11 Fraction of new behavioral and physical sciences Ph.D.s with definite commitments, 1975-1992. See Appendix Table F-20.

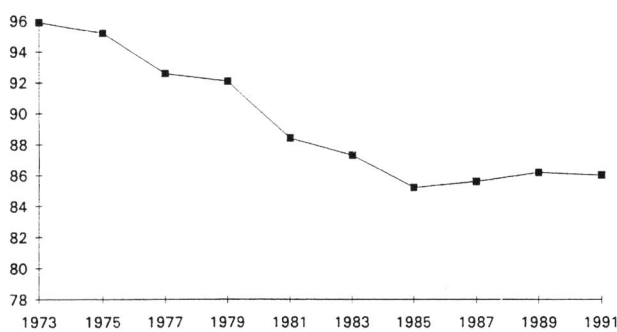

FIGURE 4-12 Salaries of behavioral science Ph.D.s (ages 30-34) who currently hold full-time employment positions (excluding postdoctoral positions) as a percentage of comparable salaries for all scientists and engineers, 1973-1991. See Appendix Table F-21.

that becomes more pronounced from 1989 to 1992 than the decline for behavioral sciences. These data suggest that the market was relatively stronger for behavioral scientists in the early part of this period but deteriorated and became relatively weak in more recent times.

Starting Salaries

Salaries are also considered to be a valid indicator of market conditions. The commonly accepted economic analysis of labor markets postulates a positive relationship between relative salaries and relative demand. This relationship is expected to be strongest at the entry point of a career, which typically occurs at around age 30 for behavioral scientists. Median salaries of employed behavioral science doctorates, age 30-34, who currently hold full-time

OUTLOOK FOR BEHAVIORAL SCIENTISTS

A major goal of this committee is to estimate future needs for behavioral scientists. As noted earlier in this report, need can be defined in a variety of ways. In the context of the labor market, need has often been expressed in terms of job openings that must be filled to attain a particular employment objective. Selection of the specific employment objective to be reached is a policy decision usually made on normative grounds. Some job openings arise from deaths, retirements, and other forms of separation from the behavioral science work force. Other job openings are generated by growth in employment opportunities. All of these job openings may be filled by recruitment from many talent pools: new doctorates, experienced doctorates from other work forces or from outside the labor market (including doctorates from abroad), nondoctorates, etc.

Given this broad context, the committee examines future employment conditions in an effort to estimate need (approximated by job openings) and our ability to meet this need (measured by new Ph.D.'s entering the behavioral sciences workforce). Because job openings can be filled by recruitment from a variety of talent pools, the reader is cautioned that the committee's indicator of our ability to meet this need represents a lower-bound estimate of this ability.

Table 4-3 shows the future number of job openings in behavioral sciences to be filled under alternative scenarios about employment growth. Three scenarios are examined: zero growth; 3.0 percent per year (the 1981-1991 compound growth rate for the behavioral science work force); and 1.5 percent per year (one-half the 1981-1991 compound rate). As noted in Chapter 3, the method used to generate these estimates is a variant of demographic cohort-survival models. It generates flows of workers into and out of this work force and among the various employment states within this work force. On the basis of these flows, it generates estimates of changes in the size and composition of this work force.[12] There are, of course, many ways to do multistate life table analysis. The data presented below should be viewed as preliminary work by the committee, which will be explored further by the Panel on Estimation Procedures in the coming months.

Average annual estimates are developed for three time periods: 1996-1997, 1998-1999, and 2000-2001. The estimates are quite sensitive to the growth rate assumption, varying from 1,404-1,512 for zero growth to 3,937-4,422 for 3.0 percent per year growth. The range is substantially narrower for a given growth rate scenario. The modest increases observed over time for a given rate of growth reflect the widely anticipated increases in deaths and retirements in the late 1990s as this work force ages. Except for the zero-growth scenario, these increases also reflect the expected overall growth in the behavioral work force.[13]

For comparison, Table 4-4 shows the number of new behavioral Ph.D.s entering the behavioral work force in previous years, estimated from the longitudinal SDR (a sample survey). Note that these numbers represent only a fraction of the degree production that occurred in these fields, because a significant number of new graduates found employment in other fields or delayed entry into the work force. An estimated 73 percent of the behavioral Ph.D.s entered this work force during the period 1985-1990.

This level of work force entry, if maintained, could more than meet the need for zero growth, but it will fall considerably short of the number needed to maintain recent growth rates.[14] It can be argued that maintenance of these recent growth rates for behavioral scientists is an unrealistic scenario. In the near future, universities are unlikely to increase faculty size dramatically, federal spending on behavioral research is not likely to increase in real terms, and private sector demand (viz., industry) is not likely to increase rapidly.

The best predictions for economic activity and R&D funding in the near future suggest that demand for behavioral scientists will grow slowly at best. Under these circumstances, maintenance of the current rate of entry of Ph.D.s in the behavioral sciences should provide an adequate supply for the years 1996-2001.

The NRSA program provides predoctoral support for nearly 600 individuals in the behavioral sciences, although only a fraction complete doctoral degrees in the same year

TABLE 4-3 Committee Estimates of the Average Annual Number of Job Openings Needed to Sustain Various Growth Rates of the Behavioral Science Work Force[a,b]

Year	Zero Growth Rate Scenario	Half the Average Growth Rate Scenario[c]	Average Growth Rate Scenario[d]
	Numbers Needed	Numbers Needed	Numbers Needed
1996-1997	1404	2592	3937
1998-1999	1441	2626	4076
2000-2001	1512	2780	4422

[a]Behavioral science work force consists of those employed or on postdoctoral appointments in a behavioral field. Data derived from the NRC Survey of Doctorate Recipients, a sample survey.
[b]Based on multistate life table methods. See Appendix G for methodology.
[c]Half the average referred to in footnote d or 1.5 percent.
[d]Refers to behavioral work force's average annual compound growth rate over the past decade or 3.0 percent (3.5 percent, uncompounded).

TABLE 4-4 Estimated Number of New Behavioral Ph.D.s Entering the Behavioral Work Force in Selected Years.

Year	Number[a]
1985-1986	2932
1987-1988	2797
1989-1990	2973

[a]Annual averages.

NOTE: The Survey of Doctoral Recipients is a sample survey and subject to sampling error.

SOURCE: NRC, Survey of Doctoral Recipients. (Biennial)

as receiving NRSA support. The number of behavioral degree recipients in any year having had NRSA support is unknown but presumed small.[15] Even if current levels of predoctoral NRSA support are maintained, then the NRSA program will most likely produce behavioral scientists at a rate which future markets will absorb.

Relationship Between Market Outlook and the Need for Behavioral Research Personnel

The preceding analysis provides background information with which we can anticipate the experiences of behavioral scientists who will join the U.S. labor force in the coming years. NRSA support although important in the behavioral sciences represents, however, a small source of research training support for new Ph.D.s in this area. Because NRSA is such a small source of support, its effect on the aggregate supply of behavioral scientists will be small. Thus, the justification for the NRSA program in behavioral sciences must be based heavily on need factors other than those generated by aggregate market conditions.

In this report, we have taken an important first step in adopting a new strategy for understanding aggregate changes in the market for behavioral scientists. We have concluded that, in the aggregate, behavioral scientists are experiencing only modest employment problems. From our global analysis of need associated with the growth in the national research agenda, we expect that NRSA award recipients in the behavioral sciences will continue to find employment. Further work is needed, however, before we reach a more satisfactory understanding of the specific needs and outlook for behavioral scientists engaged in health-related research in areas of special interest—the target group of the NRSA program. Moreover, in order to assess the impact of the program on NRSA awardees, career outcomes studies are needed.[16]

Priority Fields

The national environment for behavioral and social sciences has been variable during the past two decades (see, for example, Gerstein et al., 1988). For example, federal obligations for basic and applied research in psychology, sociology, and anthropology consistently declined throughout most of the 1970s, reaching a low of $286 million in 1982.[17] After that time it began to slowly inch back upwards until 8 years later, when it totaled $445 million (NSB, 1991). From 1985 to 1990, funds for basic research increased, on average, by $8.4 million per year, rising from $162 million to $204 million. The average annual increase in the applied research during that time was another $7 million, increasing its worth from $206 million to $241 million. Psychology has remained the primary beneficiary of federal support, accounting for 76 percent ($340 million) of the $445 million obligated in 1990, with commitments to sociology and anthropology of $93 million and $12 million, respectively.

Societal Problems

Despite the severe funding setbacks in the early 1980s, there is now increasing recognition that many of the worst problems facing this country are primarily behavioral in character and that these sciences possess important information to address those problems. The information that has accumulated over years of careful basic research is adequate to design developmental programs that have reasonable prospects for success.

A key link in relating basic findings from the behavioral sciences to a wide variety of clinical and applied settings is the recruitment and training of high-quality people to carry out this linkage. Perhaps the single largest potential for doing this lies in the NIH pre- and postdoctoral fellowship program. Were these to be viewed by NIH as a major activity, surely university-medical school consortiums would design fellowship opportunities which would produce scientists well-trained both in aspects of the behavioral sciences and in the problems encountered in medical settings.

NIH (as reconfigured in October 1992) has for many years offered multidisciplinary training opportunities in research related to the solution of such pressing social problems as substance abuse, violence, and the prevention of infectious diseases. However, because of competing but important developments in the biological bases of behavior, NIH has directed less attention to research training in the behavioral and social dimensions of physical and mental health than might be desired. It is the hope of the committee that expanded NRSA support in the behavioral sciences in the coming years will result in significant growth in the number of awards made for research training in the solution of social problems related to the health and well-being of all Americans.

A Special Note on Clinical and Clinical Services Research

Consideration should be given to enhancing federal support for training programs to produce clinical investigators (see Kraut, 1993, for suggestions). Currently, some monies are available to support mental health services research training, but few opportunities exist for supporting the production of behavioral science investigators in other problem areas (e.g., alcohol and drug abuse interventions and psychotherapy outcomes). Most clinical research training being sponsored by the National Institute of Mental Health (NIMH), for example, is primarily oriented toward psychiatry; of the 35 clinical research training grants this year, only 5 are in psychology departments, with the remainder in psychiatry units. As a result of NIMH's reorganization in 1985, clinical training programs were relocated from the research divisions to the services component and now, with the movement of NIMH to NIH, are housed in the Center for Mental Health Services (CMHS) in SAMHSA. Although subsequent engagement in research fulfills the payback requirement for clinical training, only a small proportion of previously supported trainees actually utilized this option (13.2 percent of all individuals engaged in payback requirements between 1981 and 1993), and a sample survey of former trainees indicated that only a small fraction (1.8 percent) cited research as their current primary activity (CMHS, 1993). Furthermore, the clinical training budget totaled only $2.9 million in 1992 and is a candidate for elimination in congressional budget hearings.

Previous recommendations regarding the interdisciplinary nature of conducting research in these areas (for both behavioral and clinical scientists), exposure to applied research settings during the training process, and methodological training are critical to promoting these areas. Clinical training programs are not always directed at research competence per se (understanding research so as to keep up with the literature vs. knowing how to do it). This is reflected by nearly half (48.6 percent) of all Ph.D.s and Psy.D.s awarded in clinical psychology in 1989 being from programs that emphasized the practitioner model rather than the scientist-practitioner model of training, up from 28.7 percent in 1979 (Pion, 1993). Even in programs that train students for research careers, the clinical training process itself, similar to medicine, is structured (of necessity) to meet the increasing demands of accreditation agencies whose aim is to ensure practice competencies but not minimal standards of research conduct. For example, the requirement of a predoctoral internship coupled with the need of postdoctoral clinical training to meet licensing requirements affects both interest in and ability to carry out research for both future clinicians and clinical faculty (many clinical training programs require their faculty to be licensed). Expanding the pool of clinical researchers can be accomplished in other ways by establishing specific programs (e.g., the program in clinical sociology at Yale) and by exposing nonclinical students in traditional disciplines to key substantive courses in prevention and treatment and research opportunities in settings where they work with clinicians on projects. For example, certain problems surrounding diagnostic accuracy and clinical decision-making can pose interesting research questions for cognitive psychologists and also lead to improvements in the application and teaching of these skills. Transmission of the human immunodeficiency virus (HIV) can be better understood by the application of social network models and ethnographic studies, and models for interagency collaboration (e.g., resource dependency theory) can be examined by organizational psychologists and sociologists. In general, the issue of how best to produce clinical investigators has recently generated much attention, along with serious examination of how accreditation practices have adversely affected the production of clinician scientists (see the Accreditation Summit held in 1992 by the American Psychological Society). Aside from the question of exactly how many clinical researchers are needed, the data suggest that current programs are less successful than expected in producing researchers who can examine key clinical problems and clinicians who can incorporate the logic of research into their service-delivery activities. Concerted efforts are needed to identify the most appropriate strategies for the development of clinical researchers (perhaps through retrospective examination of previous trainees and pilot tests of innovative programs).[18] Depending on the types of changes resulting from national health care reform (e.g., the changes in service delivery settings, practitioner responsibilities, and practitioner autonomy as a result of managed care), opportunities for clinical and clinical services research in these settings may expand along with the attractiveness of research careers.

ENSURING THE DIVERSITY OF HUMAN RESOURCES

The characteristics of the graduate student population have changed significantly since the 1960s. Thirty years ago the flow of women into and through doctoral programs, regardless of discipline, was exceedingly modest, with only 10.7 percent of the 9,733 doctorates awarded in the United States earned by women. Beginning in the early 1970s, however, the number and proportion of women doctorate recipients rose dramatically. By 1991 it had more than tripled, to 13,765, accounting for 36.8 percent of all new doctorate recipients (Ries and Thurgood, 1993). This marked growth occurred in nearly every broad field of inquiry, particularly the behavioral sciences.

This pattern, however, has been a product of differential trends in gender participation across disciplines. First, there was a dramatic upsurge in the number of women earning doctoral degrees, which was a product of the substantial growth in the pool of women eligible to earn the doctorate in all fields. Second, the pool of eligible men has remained relatively stable or undergone some erosion, and this is reflected in the number being awarded doctoral degrees, which has been steadily shrinking. Whereas the same percentage of women baccalaureates (approximately 3 percent) has gone on successfully to earn the doctorate, the proportion of men who enter and complete doctoral programs dropped from 10 percent to 4 percent during this same period. Consequently, the flow of women into the graduate training pipeline has increased, but the stream of men subsided from its early-1970s level.

Other factors aside (e.g., the quality of baccalaureate training), the recent upsurge in the awarding of baccalaureate degrees suggests a larger pool of candidates who may be potentially interested in pursuing advanced degrees in these areas. It also suggests that if current patterns persist, the composition of the graduate student population will change accordingly (see following section). Among current freshmen, interest in the social and behavioral sciences has increased, with much of the growth attributable to females. In 1990, 55 percent of all freshmen intending to major in a science or engineering field chose psychology or one of the social sciences as compared with 39 percent in 1983 (NSF, 1991). Given that freshmen intentions have signaled trends in baccalaureate production, graduate training programs can expect the pool of prospective applicants to have, at least in the short term, a large proportion of women, and although more ethnically diverse than many other science and engineering fields, to remain sparsely populated by people of color. Ethnic minority representation has remained relatively stable, with about 1 of every 10 graduate students being Black, Asian, Native American, or Hispanic. The recruitment and retention of ethnic minorities has been and remains an important concern.

In addition, the long-term implications of shifting gender distributions have come under review (e.g., APA Task Force on the Changing Gender Composition of Psychology), with available research suggesting that differences exist in career tracks and advancement. Although often each difference is reasonably small, the results are cumulative advantages and disadvantages that appear gender-specific (e.g., Hornig, 1987; Long, 1990; Zuckerman, 1987). Such disparities do not work toward maintaining the health and vitality of the investigator pool, and well-designed research studies investigating the antecedent factors for choosing and pursuing a research career are urgently needed.

THE NRSA PROGRAM IN THE BEHAVIORAL SCIENCES

For operational purposes, the committee created—and first reported in 1975—a tentative taxonomy of behavioral science research fields using the departmental classification scheme of the Doctorate Records File maintained by the NRC. Throughout its tenure the committee continued to define the behavioral sciences as psychology, anthropology, sociology, and speech and hearing sciences because these were the fields most closely involved in investigating health problems.

In its earliest reports, the committee presented the labor market outlook for Ph.D.s in the behavioral sciences as a whole, without distinguishing fields. In 1978 the committee, realizing that the analysis of the labor market was hindered by treating the behavioral sciences as a single entity, separated the data into clinical (clinical psychology, counseling and guidance, and school psychology) and nonclinical (nonclinical psychology, anthropology, sociology, and speech and hearing sciences) fields. This disaggregation enabled the identification of divergent market trends within the behavioral sciences.

For its 1985 report the committee decided to carry the disaggregation one step further and divided the nonclinical fields into nonclinical psychology and other behavioral sciences. This step yielded three behavioral science subdivisions: clinical psychology, nonclinical psychology, and other behavioral science fields (sociology, anthropology, and speech pathology and audiology).

With this additional level of disaggregation, substantial differences among disciplines and education levels (graduate and undergraduate) began to emerge in the 1980s. For example:

- undergraduate enrollments in psychology were not subject to the decline experienced by the other behavioral fields;
- graduate enrollments in psychology declined less than those in other behavioral science fields;
- the decline in R&D funding occurred in behavioral fields outside of psychology;
- academic employment continued to grow, but the increases were concentrated in the fields outside of psychology: sociology, anthropology, and speech pathology and audiology; and
- although the number of behavioral scientists on postdoctoral appointments rose, the number of nonclinical psychologists with postdoctoral appointments fell to its lowest level since the committee began monitoring these data.

In general, the committee's recommendations for behavioral sciences personnel were consistent throughout the 1970s with respect to the numbers of trainees and fellows supported, level of dollars expended, and distribution of awards between predoctoral levels. The core of the recommendations was, within the framework of a constant dollar level of support, to redistribute awards in the behavioral sciences from the traditional predoctoral emphasis (70 percent/30 percent, predoctoral to postdoctoral) to one of postdoctoral emphasis (70 percent/30 percent, postdoctoral to predoctoral). The committee recommended that this transition be completed by fiscal year 1981. The rationale for the recommendation was that the less favorable labor market for behavioral scientists, especially in academia, dictated a decrease in predoctoral support whereas the growing sophistication of behavioral research in the area of health warranted an increase in postdoctoral support.

During the early 1980s the committee reported that a shift in emphasis to postdoctoral training awards had occurred, but not in the gradual and orderly manner recommended. Instead, the ratio had moved slowly toward a higher postdoctoral concentration because of a sharp decline in predoctoral awards and a level of postdoctoral support that had risen between 1975 and 1977 and then remained basically constant through 1981.

Although the ADAMHA had supported the committee's recommendations and made strong efforts to implement them, significant external impediments had prevented ADAMHA's ability to respond more rapidly: overall funds available for research training grants and fellowships had lost ground to inflation, mandated increases in stipend levels in fiscal year 1980 meant that fewer pre- and postdoctoral students could be supported, and professional institutions had been slow in broadening their emphasis to include more postdoctoral training positions.

The committee in its 1983 report recommended that despite some necessary budgetary reductions, some training support must be preserved in the health-related behavioral and social sciences and directed to the highest quality programs. The recommended shift in emphasis from predoctoral to postdoctoral awards was reaffirmed, but some numerical modifications were necessary because of budgetary constraints. The committee recommended that predoctoral training should remain at the 1981 level (about 650 awards) whereas postdoctoral support should increase only modestly from the 1981 level (about 350 awards) to about 540 awards by 1987.

By the mid 1980s the committee was reporting an academic labor market for behavioral science Ph.D.s that was expanding at a moderate rate of growth. However, most of the growth had occurred in the field of clinical psychology. This finding caused the committee concern from the standpoint of the federal government's research program because most clinical psychologists work outside the academic sector and do not contribute to the research effort. Also of concern to the committee was the continuing decline of R&D funding and the observation that graduate enrollments and Ph.D. production in nonclinical fields were beginning to decline whereas those in the clinical fields were continuing to increase. However, the committee noted that a substantial number of behavioral science courses were being taught to graduate students at professional schools—public health, law, medicine, nursing, social work, and business—which would tend to increase the demand for behavioral scientists.

Recommendations in 1985 were for 450-790 predoctoral trainees per year and 460-800 postdoctoral trainees per year. The committee also strongly endorsed the continuation of the training grant mechanisms as a way of improving the quality of graduate education in the behavioral sciences at the predoctoral and postdoctoral levels. It was recommended that training grants should be the predominant mechanism of support, with an 80 percent/20 percent ratio of traineeships to fellowships.

Finally, in its 1989 report, the committee projected that the labor market for behavioral scientists would be fairly stable. It recommended that the level of predoctoral and postdoctoral support be kept at their current levels of approximately 500 and 420 positions, respectively. Given the low level of research involvement by clinical psychologists, the committee recommended moving support away from clinical psychology and toward nonclinical psychology and other behavioral sciences.

RECOMMENDATIONS

Total support for research training in the behavioral sciences increased slightly from about 902 awards in fiscal 1991 to an estimated 1,069 awards in fiscal 1993 (Table 4-5). Most awards are offered as institutional training grants

TABLE 4-5 Aggregated Numbers of NRSA Supported Trainees and Fellows in Behavioral Sciences for FY 1991 through FY 1993

Fiscal Year	Level of Training	TOTAL	Type of Support	
			Traineeship	Fellowship
1991	Number of awards	902	775	127
	Predoctoral	519	472	47
	Postdoctoral	338	258	80
	MARC Undergraduate	45	45	-
1992	Number of awards	908	790	118
	Predoctoral	534	481	53
	Postdoctoral	323	258	65
	MARC Undergraduate	51	51	-
1993	Number of awards	1,069	930	139
	Predoctoral	672	604	68
	Postdoctoral	349	278	71
	MARC Undergraduate	48	48	-

NOTE: Based on estimates provided by the National Institutes of Health. See Summary Table 1.

(traineeships), which account for about 85 percent of total NRSA support in the behavioral sciences. Emphasis has been given to predoctoral support, although institutional training grants permit the mix of predoctoral and postdoctoral trainees. Individual fellowships have been made available at both the predoctoral and postdoctoral levels.

Predoctoral Training

On the basis of continuing gains being made by behavioral scientists in areas of national interest and of anticipated employment opportunities for highly skilled researchers, the committee urges the continued expansion of federal support through predoctoral awards in the behavioral sciences. Predoctoral awards permit the preparation of investigators familiar with the broad range of research techniques and theories that characterize doctoral preparation in the behavioral sciences. Many graduates are ready to assume research positions on completion of the doctoral degree, although postdoctoral training has gained some momentum in certain component subfields.

As is the case in other areas, the committee is concerned that current low stipend levels for NRSA awardees serve as a disincentive to attract the most able scientists to research careers in health-related fields. Thus, the committee has tempered its call for expansion in total support in the behavioral sciences in recognition of the competing need to increase stipend support (Table 4-6).

RECOMMENDATION: The committee recommends that the number of predoctoral trainees and fellows supported annually in the behavioral sciences increase from an estimated 672 in fiscal 1993 to 900 by fiscal 1996. The expansion in support should maintain the same ratio of trainees to fellows.

Postdoctoral Training

Postdoctoral research training through the NRSA award provides the nation with a mechanism to attract the most skilled scientists to address areas of national need. Because of differences in the evolution of research careers, postdoctoral research training plays a greater role in some behavioral sciences than others. Nonetheless, postdoctoral studies increase the technical skills of the investigator and strengthen the pool of talent available to the nation for research.

RECOMMENDATION: The committee recommends that the number of postdoctoral trainees and fellows supported annually in the behavioral sciences increase from approximately 349 awardees in fiscal 1993 to 500 in fiscal 1996.

Minority Access to Research Careers

Since its inception in the late 1970s, the special program of undergraduate support for Minority Access to Research

TABLE 4-6 Committee Recommendations for Relative Distribution of Predoctoral and Postdoctoral Traineeship and Fellowship Awards in Behavioral Sciences for FY 1994 through FY 1999

Fiscal Year	Level of Training	TOTAL	Type of Support	
			Traineeship	Fellowship
1994	Recommended number of awards	1,195	1,040	155
	Predoctoral	745	670	75
	Postdoctoral	400	320	80
	MARC Undergraduate	50	50	-
1995	Recommended number of awards	1,325	1,150	175
	Predoctoral	825	740	85
	Postdoctoral	450	360	90
	MARC Undergraduate	50	50	-
1996	Recommended number of awards	1,450	1,260	190
	Predoctoral	900	810	90
	Postdoctoral	500	400	100
	MARC Undergraduate	50	50	-
1997	Recommended number of awards	1,450	1,260	190
	Predoctoral	900	810	90
	Postdoctoral	500	400	100
	MARC Undergraduate	50	50	-
1998	Recommended number of awards	1,450	1,260	190
	Predoctoral	900	810	90
	Postdoctoral	500	400	100
	MARC Undergraduate	50	50	-
1999	Recommended number of awards	1,450	1,260	190
	Predoctoral	900	810	90
	Postdoctoral	500	400	100
	MARC Undergraduate	50	50	-

Careers (MARC) has included a small number of awards for research training in the behavioral sciences. In recent years, about 50 awards were made for training in this area (Table 4-5).

The MARC program represents a unique and special program of support for the recruitment of minorities into research careers in the behavioral sciences. We look forward to reviewing the outcome of the review of the MARC program presently being conducted under the auspices of NIH. We endorse the continuation of MARC support for research training in the behavioral sciences at current levels until that study is complete.

RECOMMENDATION: The committee recommends that the number of NRSA awards for research training in the behavioral sciences through the MARC program be maintained at about 50 per year, pending completion of the NIH review.

NOTES

1. Broadly construed to include psychology, sociology, anthropology, and speech and hearing sciences. See Appendix B for a list of disciplines included in this area.

2. There was a significant drop in the academic sector work force between 1989 and 1991. In part, this decline may reflect methodological changes that were introduced at that time to the Survey from which these data were generated. However, the decline may be reflecting a weakening of demand in the academic sector.

3. Approximately 30,000 thousand of the 67,000 workers employed in the behavioral sciences in 1991 were clinical psychologists whose participation in health-related research varies from year to year.

4. See Appendix F.

5. Survey of Doctoral Recipients (SDR) data, special tabulations.

6. See Appendix F.

7. In 1989 hospitals and other nonprofit institutions employed 23 percent of the clinical psychologists, seven percent of the nonclinical psychologists, and five percent of the sociologists and anthropologists (Pion, 1993). The numbers in Pion's paper come from Quantum Research Corporation's analysis of the SDR data. Those numbers are not strictly comparable with labor force numbers used elsewhere in this report, due to differences in definition of variables.

8. See Chapter 3, Reis, P. and D.H. Thurgood, *Summary Report 1992: Doctorate Recipients from United States Universities*, Appendix Table B-1.

9. Although the number of postdoctoral appointments are also thought to be strongly influenced by current job opportunities, we do not include them as an indicator for behavioral scientists because, as shown earlier, they are a relatively unimportant part of this labor market.

10. Starting salaries are defined as the salaries of workers in the age group 30-34.

11. See IOM, 1994 for further discussion of this phenomenon.

12. For more detail, Chapter 3, note 11 and Appendix G.

13. The increases could also be reflecting the impact of the expected aging of this work force on separations other than deaths and retirements. See supra, Chapter 3, note 9.

14. The estimate is presented as a minimum value because these job openings could also be filled by recruiting workers with degrees and training in closely related fields or workers from abroad.

15. On the basis of information gathered from the National Science Foundation the committee estimates that fewer than 5 percent of graduate students in the behavioral sciences received NRSA support in FY 1990.

16. See Appendix A for a brief review of earlier NRC studies in this area.

17. These figures are in constant 1982 dollars.

18. Some lessons can be learned from the training of physician-scientists and the performance of trainees who are supported during predoctoral years (M.D.-Ph.D. programs) vs. postdoctoral support on research-related outcomes.

REFERENCES

Adler, N. and K. Matthews
　1994　Health Psychology: Why do some people get sick and some stay well? In *Annual Review of Psychology*, Volume 45, ed. L. Porter and M. Rosenzweig, pp. 229-259. Palo Alto, CA: Annual Reviews Inc.

Bachevalier, J.
　1991　Cortical versus limbic immaturity: Relationship to infantile amnesia. In *Developmental Behavioral Neuroscience*, ed. M.R. Gunner and C.A. Nelson, pp. 129-153. Hillsdale, NJ: Lawrence Erlbaum Associates, Publishers.

Beardsley, G. and M.G. Goldstein
　1993　Psychological factors affecting physical condition. Endocrine disease literature review. *Psychosomatics*. 34(1):12-19.

Brown, J.D. and K.L. McGill
　1989　The cost of good fortune: When positive life events produce negative health consequences. *Journal of Personality and Social Psychology*. 57:1103-1110.

Center for Mental Health Services (CMHS)
　1993　*Report to Congress on Mental Health Clinical Training Program Payback Requirements*. June. Rockville, MD: Substance Abuse and Mental Health Services Administration (SAMHSA).

Clark, G.A. and E.M. Schuman
　1992　Snails' tales: Initial comparisons of synaptic plasticity underlying learning in Hermissenda and Aplysia. In *Neuropsychology of Memory*, 2nd Edition, ed. L.W. Squire and N. Butters. New York: Guilford Press.

Consortium of Social Science Associations (COSSA)
　1994　*Washington Update*. 13(4):18. March 7.

Costa, P.T., Jr. and R.R. McCrae
　1987　Neuroticism, somatic complaints, and disease: Is the bark worse than the bite? *Journal of Personality*. 55:299-316.

Cox, D.J. and L. Gonder-Frederick
　1992　Major developments in behavioral diabetes research. *Journal of Consulting and Clinical Psychology*. 60:628-638.

Fitzgerald, K.
　1989　Medical electronics. *IEEE Spectrum*. 26:67-69.

Gerstein, D., R.D. Luce, and N.J. Smelser
　1988　*The Behavioral and Social Sciences: Achievements and Opportunities*. Washington, D.C.: National Academy Press.

Getty, D.J., R.M. Pickett, C.J. D'Orsi, and J.A. Swets
　1988　Enhanced interpretation of diagnostic images. *Investigative Radiology*. 23:240-252.

Green, D.M. and J.A. Swets
　1966　*Signal Detection Theory and Psychophysics*. New York: Robert E. Krieger Publishing Company.

Goodwin, F.K.
　1993　NIH adds institutes on mental and addictive disorders. *NIH News and Features*. Summer: 12-25.

Hahn, R.C. and D.B. Petitti
　1988　Minnesota Multiphasic Personality Inventory-rated depression and the incidence of breast cancer. *Cancer*. 61:845-848.

Hitchcock, J.M. and M. Davis
　1986　Lesions of the amygdala, but not of the cerebellum or red nucleus, block conditioned fear as measured with the potentiated startle paradigm. *Behavioral Neuroscience*. 100:11-22.

Hornig, L.
　1987　Women Graduate Students. In *Women: Their Underrepresentation and Career Differentials in Science and Engineering*, ed. L.S. Dix, pp. 103-122. Washington, D.C.: National Academy Press.

Institute of Medicine
　1994　*Careers in Clinical Research: Obstacles and Opportunities*. Washington, D.C.: National Academy Press.

Kandel, E.R.
　1976　*Cellular Basis of Behavior*. San Francisco: W.H. Freeman and Company.

Kraut, A.
　In Press　Presentation at the Committee Public Hearing, May 3, 1993. In *Meeting the Nation's Needs for Biomedical and Behavioral Scientists: Summary of the 1993 Public Hearing*. Washington, D.C.: National Academy Press.

Loftus, G.R. and E.F. Loftus
　1976　*Human Memory: The Processing of Information*. Hillsdale, NJ: Lawrence Erlbaum Associates, Publishers.

Long, J.S.
　1990　The Origins of Sex Differences in Science. *Social Forces*. 68(4):1297-1315.

Luce, R.D., N.J. Smelser, and D.R. Gerstein
　1989　*Leading Edges in Social and Behavioral Science*. New York: Russell Sage Foundation.

Markovitz, J.H., K.A. Matthews, R.R. Wing, L.H. Kuller, and E.N. Meilahn
　1991　Psychological, biological, and health behavior predictors of blood pressure change in middle-aged women. *Journal of Hypertension*. 9:399-406.

Matthews, K. A., K.L. Woodall, T.O. Engebretson, B.S. McCann, and C.M. Stoney
- 1992 Influence of age, sex, and family on Type A and hostile attitudes and behaviors. *Health Psychology.* 11:317-323.

McGaugh, J.
- 1989 Involvement of hormone and neuromodulatory systems in the regulation of memory storage. *Annual Review of Neuroscience.* 12:255-287.

McNeil, B.J., S.G. Pauker, H.C. Sox, Jr., and A. Tversky
- 1982 On the elicitation of preference for alternative therapies. *New England Journal of Medicine.* 306:1259-1262.

Metz, C.E.
- 1986 ROC methodology in radiologic imaging. *Investigative Radiology.* 21:720-733.

Meuner, M., J. Bachevalier, M. Mishkin, and E.A. Murray
- 1993 Effects on visual recognition of combined and separate ablations of the entorhinal perirhinal cortex in rhesus monkeys. *Journal of Neuroscience.* 13(12):5418-5432.

National Institutes of Health (NIH)
- 1991 *Health and Behavior Research: NIH Report to Congress.* Bethesda, MD: National Institutes of Health.
- 1993 *The NIH Implementation Plan for Health and Behavior Research.* Bethesda, MD: National Institutes of Health.

National Science Board (NSB)
- 1991 *Science and Engineering Indicators - 1991.* Washington, D.C.: National Science Foundation.

O'Keefe, J. and L. Nadel
- 1978 *The Hippocampus as a Cognitive Map.* London: Oxford University Press.

Pion, G.
- 1993 Trends in behavioral science human resources: their implications for research training. Paper prepared for the Committee on National Needs for Biomedical and Behavioral Research Personnel.

Ries, P. and D.H. Thurgood
- 1993 *Summary Report 1992: Doctorate Recipients from United States Universities.* Washington, D.C.: National Academy Press.

Salovey, P. and D. Birnbaum
- 1989 Influence of mood on health-relevant cognitions. *Journal of Personality and Social Psychology.* 57:539-51.

Scheier, M. F. and C.S. Carver
- 1992 Effects of optimism on psychological and physical well-being: theoretical overview and empirical update. *Cognitive Therapy and Research.* 16:201-228.

Siegrist, J., R. Peter, A. Junge, P. Cremer, and D. Seidel
- 1990 Low status control, high effort at work and ischemic heart disease: prospective evidence from blue-collar men. *Social Science and Medicine.* 31:1127-1134.

Squire, L.R.
- 1992 Memory and the hippocampus: a synthesis from findings with rats, monkeys, and humans. *Psychological Review.* 99:195-231.

Swets, J.A.
- 1979 ROC analysis applied to the evaluation of medical imaging techniques. *Investigative Radiology.* 14:109-121.
- 1988 Measuring the accuracy of diagnostic systems. *Science.* 240:1285-1293.

Swets, J.A., D.J. Getty, R.M. Pickett, C.J. D'Orsi, S.E. Seltzer, and B.J. McNeil
- 1991 Enhancing and evaluating diagnostic accuracy. *Medical Decision Making.* 11:9-18.

Swets, J.A. and R.M. Pickett
- 1982 *Evaluation of Diagnostic Systems: Methods from Signal Detection Theory.* New York: Academic Press.

Thompson, R.F.
- 1986 The neurobiology of learning and memory. *Science.* 223:941-947.

Zuckerman, H.
- 1987 Persistence and Change in the Careers of Men and Women Scientists and Engineers. In *Women: Their Underrepresentation and Career Differentials in Science and Engineering*, ed. L.S. Dix, pp. 127-156. Washington, D.C.: National Academy Press.

CHAPTER FIVE

PHYSICIAN-SCIENTISTS

Training in the clinical sciences is critical to maintaining our country's leadership in the translation of basic discoveries to meaningful patient care. As the nation demands more primary care of our physicians, we must not lose sight of the tremendous advances that have been made by individuals using basic approaches to explore interesting and significant clinical problems.

Clinical research includes a spectrum of investigation. At one end, it is represented by the use of basic scientific approaches and tissue samples from patients or normal individuals to generate fundamental insights into the disease process. At the other end, it is represented by studies in which whole patients, normal volunteers, or populations of subjects each serve as the laboratory.[1]

The clinical investigator generally has an M.D. or other health professional doctorate, although the committee recognizes that basic scientists also participate in clinical investigation. The committee has based its assessment of national need on the fact that most government-sponsored research in the clinical sciences is performed in medical schools or academic health centers (Appendix Table F-22). The ability of medical schools to conduct clinical research depends largely on the continuing availability of clinical faculty with strong research skills. The future availability of well-prepared clinical research faculty has come into question by a number of authors (Ahrens, 1992; Fredrickson, 1993; IOM, 1994). Given continuing national concern over the future supply of skilled clinical investigators, we have restricted our analyses in this chapter to the need for physician-scientists.[2]

Previous National Research Council (NRC) study committees have focused on the special role that the physician-scientist has played in bringing clinical insights to bear in the laboratory and in translating new knowledge into the context of medical practice (NRC, 1981). Almost all NRC committees that have addressed research training needs in the clinical sciences have observed that there continues to be a shortage of physicians willing to prepare for research careers. Many committees have focused on the very real effects of competing—and more lucrative—opportunities available in private practice as a reason for this trend (NRC, 1978). More recently, some committees have observed that changes occurring in the way medical schools finance their operations and structure their faculties simply does not provide an environment conducive to preparation for a research career (NRC, 1985). We concur and provide evidence elsewhere in this chapter suggesting that upcoming changes in the national support for health research and health care reform may further erode research and research-training opportunities in academic health centers.

In addition to these contextual variables, we believe the nature and timing of National Research Service Award support may directly effect the success of recruiting physicians into research careers. On the basis of information gathered by the National Institutes of Health (NIH), we believe that the Medical Scientist Training Program (MSTP) may be especially effective in launching individuals into research careers. This program was established in 1964 to permit individuals to pursue the M.D. and the Ph.D. degrees concurrently. The MSTP program has consistently had a high proportion of graduates involved in research and actively contributing to the advancement of the clinical sciences. (See also, Appendix A for a summary of available outcome studies.)

Opportunities for careers in clinical research abound. Our continuing challenge is to stimulate interest of clinicians in contributing to that effort, and the NRSA program can clearly play a role relative to that goal.[3] In the sections that follow, we will review some of the more exciting advances in clinical science that create the need for a continuing research effort, summarize the current market for these scientists, and recommend specific changes in the NRSA program that may be effective in expanding the cadre of physician-scientists needed at this time.

ADVANCES IN CLINICAL SCIENCE

Advances in clinical science have been enormous and include, but are not limited, to the following:

• Identification of the genetic defect in various genetic disorders, including cystic fibrosis. Cystic fibrosis is the most common genetic disorder in Caucasians, affecting 1 of every 2,000 children. The disease is characterized by pulmonary infections and pancreatic insufficiency and is due to a cellular defect in the development of secretions. The genetic defect associated with the disorder is found in chromosome 7. This discovery allows three major advances. First, it allows genetic counseling within families. Second, it has allowed a determination of the product of the gene. This information will provide a rational approach to developing drugs to correct the defect. Finally, it will allow studies that attempt to replace the defective gene with a normal one in tissues that are affected. Indeed, such somatic gene therapy has already begun.

• Identification of the gene associated with bowel cancer. Very recently, two separate groups of investigators demonstrated a genetic defect localized to chromosome 2, which is associated with hereditary nonpolyposis colon cancer. The gene involved appears to control DNA repair, and a defective gene seen in patients with colon cancer leads to instability of cellular DNA. This research is a spectacular example of the different ways in which basic research can lead to clinical advances. In one laboratory the research developed from studies performed in yeast and bacteria that examined how these organisms repair DNA and the genetic defects associated with DNA instability. In another laboratory there is a long history of studies in humans examining genetic defects associated with a variety of colon cancer syndromes. In other words, this remarkable advance in our understanding of colon cancer came from distinct pathways, one originating from basic studies of normal mechanisms in bacteria and yeast and the other from more clinically oriented studies looking at abnormal growth and differentiation of colon cells. These studies will allow the development of reagents that can be used to screen for colon cancer.

• Creation of an animal model for ankylosis spondylitis by using transgene methodology. Ankylosis spondylitis is a syndrome that predominantly affects joints of the spine. Approximately two decades ago it could be demonstrated that the disease was significantly associated with a specific HLA type, HLA-B27. Indeed, 90 percent of patients with ankylosis spondylitis had the HLA-B27 genotype. In an attempt to demonstrate the nature of the association between the gene and the disease, investigators established a rat model in which the human HLA-B27 gene was inserted by using transgene methodology. In some of the animals a disease developed that mimicked human ankylosis spondylitis. These animals not only provide a model for determining just how the gene influences the expression of the disease but also for deciding what other factors may be involved. They also provide a model for studying the effectiveness of various forms of therapy.

The importance of clinical research to advancing our understanding of clinical disorders is captured in a recent editorial in *Science* written by Editor-in-Chief Daniel E. Koshland Jr. (1993):

> In the 1980s and 1990s NIH researchers, intramural and extramural, performed the first trial of gene therapy in humans, proved the effectiveness of methotrexate for treating rheumatoid arthritis, developed new methods for growing skin to repair burns, showed that control of glucose levels slows progression of diabetes, showed effectiveness of cholesterol reduction in the prevention of heart disease, demonstrated an effective treatment for spinal cord injury, found a new drug for Parkinson's disease, showed that aspirin and coumadin lower the risk of stroke, developed methods of hypertension control that have reduced heart attacks and strokes by more than 50 percent, and so on for many other discoveries. .... These followed many earlier discoveries, including the polio vaccine, the measles vaccine, hormone replacement therapy, fluoride to prevent tooth decay, to name a few. We are living longer, we are living with less pain, we are living with less cost to alleviate health deficiencies than any previous generation because of the findings of health researchers. .... In the not-so-distant past, smallpox epidemics killed 25 percent of the inhabitants of towns that were invaded by the virus. Today we are storing the last traces of the virum because that dread disease has been eradicated from the Earth.

Clearly, this partial list of clinically relevant discoveries supports the practical value of clinical research. The United States is the world leader in clinical research and we must make a renewed commitment to retain this leadership. The recommendations of this report should allow us to remain in this position of preeminence.

ASSESSMENT OF THE CURRENT MARKET FOR CLINICAL SCIENTISTS

Degree Production and Career Patterns

Clinical scientists work in a variety of settings but primarily in academic health centers. Between 1981 and 1991, the number of full-time faculty employed in clinical departments grew by about 38,000 to just over 59,000 (Appendix Table F-23), suggesting that the market for clinical scientists remained relatively robust throughout the 1980s.

Degree Production

The major source of new physician-scientists is the nation's medical schools. The most readily available information about patterns of enrollment and degree production is the Association of American Medical Colleges (AAMC). Data from AAMC (Appendix Table F-23) reveals that medical school enrollments remained essentially flat between 1981 and 1991 at about 65,000 students. The number of medical degrees awarded each year also remained level at about 15,500 per year in the 1980s.

Career Patterns

Few data sets are available for sorting out the unique patterns of research careers among physician-scientists. The American Medical Association provides information about the number of physicians primarily engaged in research (Appendix Table F-23)—which fluctuated between 16,000 and 23,000 between 1981 and 1990. But these figures may also include trainees in graduate medical programs. Perhaps more telling is the trend in success rates of M.D.s who apply to the National Institutes of Health—which peaked at about 45 percent in 1987 and has leveled off at about 37 percent (on average) thereafter (Appendix Table F-23).

Market Forces

There are several influences on the availability of careers in clinical research. These influences, called market forces, range from how we have traditionally trained clinical researchers to outside industrial and governmental spending trends. As the nation begins to develop a new system of health care delivery, these market forces will become increasingly important.

Academic Health Centers

An academic health center can be defined as a medical school working in conjunction with a teaching hospital and at least one other health professional school to achieve mutually agreed upon goals for education, research, and provision of clinical care. Approximately 68 percent of NIH R01 support goes to these academic health centers. Academic health centers therefore constitute the major sites at which health-related research and research training are carried out. Moreover, a significant amount of their support for research is derived from income obtained for the provision of clinical care. This income stream is threatened by changes in healthcare reform that place academic health centers at a disadvantage with regard to the cost of providing medical care. This presents a threat to the market not only for training but also for support of trained investigators.

Pharmaceutical and Biotechnology Industry

Uncertainties in health care reform has forced industry to be exceedingly cautious with expenditures, and in some cases to lay off large numbers of employees. This posture clearly stifles innovation. One of the first areas to feel the effects of budgetary constraints is research. This soft side of the market has to be balanced by the fact that there are tremendous opportunities for the development of unique agents to treat significant clinical disorders.

Cap on Domestic Spending

The federal deficit, budget reconciliation, and a cap on domestic spending indicates that support for research and training will have to compete for many other high-priority areas supported by the domestic budget. This scenario is one in which the NIH budget is likely to grow at a rate certainly not greater and probably somewhat less than the biomedical research price index.

Emphasis on Increasing the Training of Generalists

There clearly is an enormous pressure nationally to increase the proportion of generalists in medicine and decrease the proportion of specialists. Heretofore, significant research training and research activity has occurred in association with specialties, particularly the medical specialties. Indeed, some view the problem in the imbalance of generalists to specialists as a result of overemphasis on research spending. This, therefore, provides a diminished enthusiasm among some to further increase funding for research or research training.

OUTLOOK FOR CLINICAL SCIENTISTS

In addition to market forces, there are factors that influence the demand for clinical scientists. These demand indicators are expenditures for clinical research and development (R&D) in medical schools; professional service income in medical schools; total revenue; budgeted vacancies in medical schools, both in clinical and basic science departments; and the clinical faculty/student ratio.

Expenditures For Clinical Research and Development

From 1985 to 1990, expenditures for clinical R&D in medical schools increased moderately. The average increase was about 13 percent per year. An estimate of the amount of support for clinical R&D in medical schools is needed to refine the model of demand for clinical faculty. An estimate of clinical R&D expenditures in medical schools was derived by using the proportion of total NIH obligations used to support clinical research. From 1969 to 1989 this proportion increased by 60 percent (Appendix Table F-22).

Since 1980 public medical schools have had higher levels of research expenditures than have private schools. This is partly due to the fast growth in the number of public schools. Clinical R&D in public schools grew at an annual rate of 7 percent since 1980 as compared with only about 4 percent per year in private schools. However, private schools remain more research intensive as indicated by research expenditures per school. Average clinical R&D expenditures were $14.6 million in private schools in 1990 compared with $10.2 million in public schools (Appendix Table F-24).

Professional Service Income

Service income generated by medical school faculty has continued to grow. From 1989 to 1990 service income generated by medical school faculty grew 14 percent and from 1990 to 1991 this income grew 13 percent (figures adjusted for inflation in 1987 dollars). This has increased as medical schools have become very successful in providing clinical care. Thus, medical schools have come to depend on the clinical income to support their research and educational missions.

Total Medical School Revenue

Service income and federal research funds contributed over half of all medical school revenues in 1991. Another large portion came from state and local government sources. Tuition contributed only about 4 percent in 1991. With an average yearly tuition increase since 1985 of about 6 percent, medical student indebtedness, as noted by several testifiers at the public hearing, may operate as a deterrent to their pursuit of research training. The rates of total revenues have grown at an average yearly rate of 14 percent since 1986.

Budgeted Faculty Vacancies

Total budgeted medical school faculty vacancies have grown at an average yearly rate of about 6 percent since 1989. The major growth of vacancies is in the clinical science departments. There has been a steady decrease of faculty vacancies in the basic science departments with a high of 801 budgeted vacancies in 1985 to the 1991 low of 643 vacancies.

Faculty/Student Ratio

Enrollments, revenue, and clinical faculty size are the basic elements in assessing personnel needs for the clinical sciences in medical schools. The ratio of clinical faculty to enrollment is largely determined by the funds available to support faculty.

Priority Fields

Clinical investigation requires practitioners to stay abreast of developments in both medicine and science, each of which is in constant acceleration and often the two do not track in parallel directions (Fredrickson, 1993). Observations from the study of patients lead to the development of hypotheses, which lead, in turn, to scientific experimentation. Interest in the patient is always paramount, but scientific experimentation runs the risk today of taking the clinical investigator away from the bedside to the clinical laboratory. Ahrens (1992), in particular, has decried the reductionist direction of clinical investigation, suggesting that patient-oriented research is seriously imperiled. We concur with Ahrens view that more emphasis should be placed on the preparation of investigators familiar with the experimental paradigms associated with patient-oriented research. At the same time we recognize that laboratory-based clinical investigation has a significant and continuing role in the national health effort. However, from our review of the literature, and on the basis of our expert judgement, we cannot help but conclude that there is indeed a dearth of individuals adequately trained to perform patient-oriented or population-based research.

With the development of new therapies and diagnostic procedures, there is an urgent need to train individuals who can carry these advances into the clinic so that their effectiveness can be measured and made available to the nation. NRSA funds, either through individual NRSA fellowships or programmatic training grants, can play an effective role in promoting the specialists that are needed.

The MSTP also represents a priority field. Established in the 1960s, this program has been especially attentive to the essential training requirements for clinical investigation. A 1992 study of graduates of the Johns Hopkins University's M.D./Ph.D. program found that all of those who had completed their training were actively involved in research: 81 percent in academia, 14 percent at research institutes, and 5 percent in the biotechnology industry (McClellan and Talalay, 1992. See also Bradford et al., 1986 and Frieden and Fox, 1991).

NIH has also analyzed information about first-time applicants for research grants (R01) and prior research training experience (Appendix Table F-25). They found that in 1989 nearly 60 percent of individuals holding joint M.D./Ph.D. degrees and applying for research support had received formal research training through support provided by NIH; this value was 47 percent for Ph.D. applicants and 42 percent for M.D. applicants. Furthermore, among first-time NIH grant recipients in 1989, 68 percent of the M.D./Ph.D. recipients had had previous NIH supported research training experience compared with 55 percent of the grant recipients holding Ph.D. degrees and 52 percent of those holding M.D. degrees (Appendix Table F-26). We conclude that continued and expanded support of the MSTP program will yield the cadre of active and successful physician-scientists so sorely needed for the national research effort today.

ENSURING THE DIVERSITY OF HUMAN RESOURCES

Issues remain regarding the recruitment of minorities to faculty and the retention of all M.D. investigators regardless of ethnicity and gender. In addition to improved recruitment, there must be specific attention given to the retention of women as clinical investigators and faculty. Extending the tenure clock and having on-site day-care are two examples of ways to facilitate their retention.

Race and Ethnicity

Medical school faculty reveal a race/ethnicity mix similar to the basic biomedical sciences (AAMC, 1992). Because most of the U.S. population will soon be a mixture of races other than white, the market will demand a more widely representative pool of researchers. About 13 percent of the faculty are members of minority groups and the largest share of these workers is Asian (Table 5-1). Table 5-1 displays the medical school faculty by rank and ethnicity: of the professors, 87.6 percent are white, 5.7 percent are Asian, and 2.4 percent declined to respond; of the associate professors, 82.5 percent are white, 7.9 percent are Asian, and 3.1 percent declined to respond; of the assistant professors, 77.5 percent are white and 8.6 percent are Asian, and information was missing on 4.8 percent; of the instructors, 72 percent are white and 9.4 percent are Asian, and information was missing on 7.4 percent. Although 13 percent of the entire faculty represent minorities, this mix is generally not yet reflected in higher faculty ranks.

Age

Figure 5-1 shows the distribution of U.S. medical school faculty by age. Out of a total of 70,187 faculty, 57.6 percent are ages 40-49 and 25 percent are ages 30-39. Table 5-2 indicates that of those aged 40-49, 22.4 percent are full professors, 55.3 percent are associate professors, 39.1 percent are assistant professors, and 29.9 percent are instructors. Of those aged 30-39, 0.3 percent are professors, 8.1 percent are associate professors, 45.8 percent are assistant professors, and 51.8 percent are instructors. Of the M.D./Ph.D. graduates, 39.8 percent are ages 40-49 and 25 percent are ages 50-59 (Table 5-3). Only 15.6 percent M.D./Ph.D. graduates are ages 30-39.

TABLE 5-1 Distribution of U.S. Medical School Faculty by Rank and Ethnicity

Ethnicity	Professor		Associate Professor		Assistant Professor		Instructor	
	Number	%	Number	%	Number	%	Number	%
Native American	24	0.1	13	0.1	22	0.1	11	0.2
Asian	1,065	5.7	1,334	7.9	2,300	8.6	590	904
Black	193	1	319	1.9	778	2.9	300	408
Mexican American	28	0.1	41	0.2	114	0.4	19	0.3
Puerto Rican	96	0.5	146	0.9	247	0.9	88	104
Other Hispanic	258	1.4	255	1.5	461	107	129	201
White	16,396	87.6	14,008	82.5	20,838	7705	4,512	72
Refused [a]	446	2.4	531	3.1	843	301	156	2.5
Missing	215	1.1	334	2	1,293	408	465	704
Total	18,721	100	16,981	100	26,896	100	6,270	100

a Declined to respond.

SOURCE: Association of American Medical Colleges (1992).

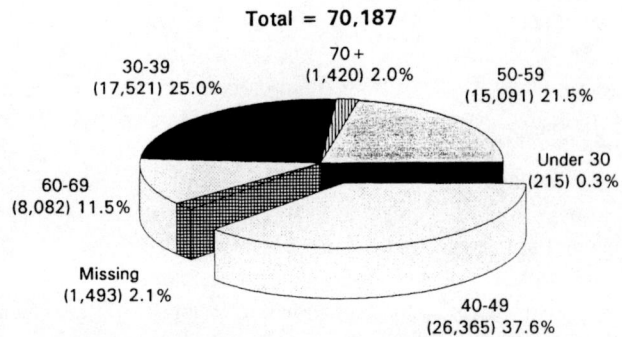

FIGURE 5-1 Distribution of U.S. medical school faculty by age.
SOURCE: Association of American Medical Colleges (1992).

THE NRSA PROGRAM IN THE CLINICAL SCIENCES

Every NRC study committee has noted that recruitment of qualified clinical researchers poses special challenges. Physicians, dentists, and veterinarians enjoy several attractive career alternatives. The vast bulk of Ph.D. trainees pursue research careers, but only one-third of postdoctoral M.D. trainees have followed that path, with most entering medical practice instead. Part of the reason is that preparing clinical specialists for practice in shortage fields has been an explicit purpose of the NIH funding programs. With that goal now substantially met, however, the committee doubted the need for continued subsidy to clinical training for practice.

TABLE 5-2 Distribution of U.S. Medical School Faculty by Degree and Age

Age	Professor		Associate Professor		Assistant Professor		Instructor	
	Number	%	Number	%	Number	%	Number	%
Under 30	0	0	0	0	58	0.2	102	1.6
30-39	62	0.3	1375	801	12314	45.8	3248	51.8
40-49	4185	22.4	9397	55.3	10512	39.1	1875	29.9
50-59	7835	41.9	4265	25.1	2286	8.5	533	8.5
60-69	5526	29.5	1532	9	790	2.9	168	2.7
70+	1038	5.5	220	1.3	129	0.5	25	0.4
Missing	75	0.4	192	1.1	807	3	319	5.1
Total	18721	100	16981	100	26896	100	6270	100

SOURCE: Association of American Medical Colleges (1992).

TABLE 5-3 Distribution of U.S. Medical School Faculty by Rank and Age

Age	M.D.		Ph.D./O.H.D.[a]		M.D.-Ph.D./M.D.-O.H.D.[a]		Other[b]	
	Number	%	Number	%	Number	%	Number	%
Under 30	60	0.1	58	0.3	1	0	96	2.1
30-39	12114	28.3	3715	19.7	600	15.6	1092	23.6
40-49	15235	35.6	7976	4202	1529	39.8	1625	35.2
50-59	8786	20.5	4652	24.6	960	25	693	15
60-69	5203	12.1	1995	10.6	586	15.3	298	6.4
70+	971	2.3	268	1.4	131	3.4	50	1.1
Missing	456	101	237	1.3	31	0.8	769	16.6
Total	42825	100	18901	100	3838	100	4623	100

a O.H.D.: Other health doctorate.
b Other: Faculty with non-doctoral/no degree or missing degree data.

SOURCE: Association of American Medical Colleges (1992).

In its first report in 1975, the committee found the available data on clinical researchers to be wholly inadequate to its needs, and it declined to recommend any change in existing levels of funding: 140 predoctoral and 3,340 postdoctoral clinical sciences trainees. Beginning with the 1976 report, the committee began to grapple in earnest with the scope of its task. On the basis of unique value and the special demands of clinical research and the fact that professional schools do not ordinarily prepare students for careers as researchers, the committee concluded that postdoctoral clinical trainees should generally receive their support in the form of training grants made to professional schools, which permit these institutions to build in a short time the critical mass of students and web of resources necessary for high-quality programs.

Evidence available in the early 1970s suggested that, unlike the burgeoning supply of Ph.D. researchers, the pool of M.D. investigators was shrinking while demand was growing steadily. Despite 6 percent annual growth in medical school faculties, the American Medical Association figures showed a significant drop since 1968 in the number of physicians engaged primarily in research. Therefore, the committee recommended funding a total of 2,800 postdoctoral traineeships and fellowships, up 10 percent from the number funded in 1975. It also praised the MSTP, initiated in fiscal year 1964, which supported students in combined 6-year M.D./Ph.D. courses. The committee recommended funding 600 MSTP traineeships, up 19 from the 581 funded in 1975.

With the next several reports beginning in 1977, the committee began exploring why the number of physicians-scientists was dropping. Although both enrollment and R&D funding were rising rapidly at medical schools, many established clinical faculty members were spending relatively little of their time conducting research. In response to this finding, the committee detailed a number of factors that it believed might discourage physicians from undertaking research careers:

- the risk of failing at an untried field after demonstrating the ability to succeed in medical practice,
- the loss of income as compared to practice,
- a growing perception among students that patient care has greater value than research,
- social pressure on students to enter primary care fields, and
- an image that paperwork and red tape inhibit researchers more than in the past.

In addition, the committee noted a discrepancy between the medical training calendar and the NIH grant cycle. Physicians who were planning for residencies to begin in July had to do so as early as the preceding October, many months before NIH announced its training awards.

In view of these circumstances, the committee continued through the 1970s to recommend 2,800 postdoctoral traineeships and fellowships in clinical sciences. It also continued to praise the MSTP awards as a highly effective method of producing clinical researchers, recommending incremental increases in the program.

By 1979 the committee's warnings appeared to have had an effect. Presidents of four leading societies discussed the threat of clinical investigator shortage in major addresses, as did a conference at the University of Chicago. Various agencies had already begun trying to counter the shortage. NIH, for example, had expended and modified its grant mechanism to ease the transition from medical school to research training and then to independent research. A 1978 amendment to the NRSA Act encouraged students to do short-term research under 3-month grants not subject to payback.

Several other developments that pointed toward a brighter outlook were an increase in the number of physicians reporting research as a major activity, an increase in the number of clinical science traineeships and fellowships, and survey results showing that medical students were growing more interested in research. The committee maintained its recommendation of 2,800 postdoctoral awards.

During the early 1980s the committee continued to recommend holding the number of awards stable at 2,800. Market opportunities for clinical investigators continued to be favorable, with medical school faculties still growing and providing places for young scientists interested in research careers. The immediate problem was the recruitment of physicians to undertake research training. The committee was concerned about a looming physician surplus, which would probably slow the growth of medical school enrollments and faculty and in turn reduce the positions available to new clinical researchers. Even with fewer openings, however, the committee believed that clinical investigator posts would remain hard to fill.

In 1985 the committee recommended a rise in the number of NRSA awards. It believed that demand would grow faster than expected, in part because of increasing attrition from an aging faculty pool.

The 1985 report also highlighted some important changes in medical school financing that the committee feared might further weaken clinical departments' commitment to research. As revenue from patient care steadily climbed, the committee believed that clinical faculty might find these demands competing for the time needed to prepare proposals, collect data, write grants, and so forth. In addition, as faculties grow less rapidly, medical school might favor hiring clinician-teachers over physician-scientists.

The committee also examined factors affecting young dentists' decision to pursue careers in clinical research. Although, unlike physicians, dentists have ample opportuni-

TABLE 5-4 Aggregated Numbers of NRSA Supported Trainees in the Medical Scientist Training Program (MSTP) for FY 1991 through FY 1993.

Fiscal Year	Number of Predoctoral Trainees
1991	783
1992	806
1993	822

NOTE: Based on estimates provided by the National Institutes of Health. See Summary Table 1.

ties for research during postdoctoral specialty training, only a few trainees receive salaries and some must even pay tuition. The committee recommended special consideration to providing adequate support for training dentist-researchers.

In 1989 the committee noted that the number of NIH traineeships and fellowships for clinical investigators (whom it chose to call physician/scientists) had not increased as fast as health-related R&D expenditures. The percentage of M.D.s who were principal investigators on NIH research grants had fallen, although the number of M.D./Ph.D. principal investigators had remained constant for the past decade.

The committee speculated that the demand for physician-scientists would increase in the future as health-related R&D increased. However, given the lack of compelling data about supply and demand and questions about the effectiveness of physician research training, the committee recommended that the number of training positions remain the same until current training programs could be evaluated.

RECOMMENDATIONS

The following recommendations are made to enhance our excellence in physician-based research.

The Medical Scientist Training Program

In 1963, NIH granted funds to three institutions to support just under 20 individuals who pursued the M.D. and Ph.D. degrees concurrently. Early NRC study committees reported findings from studies that consistently showed that a substantial fraction of MSTP awardees remain productively engaged in research, often with greater success in securing research support than M.D.s who pursue post-M.D. research training not leading to a doctorate.

Current support for M.D./Ph.D. training provides for about 820 awards (Table 5-4). Given the success of this program in contributing workers to the national research effort and the continuing and increasingly difficult problem of attracting M.D.s without Ph.D. training to research careers, we believe this program should be expanded significantly in the coming years (Table 5-5).

RECOMMENDATION: To meet the nation's continuing need for clinical investigators, the committee recommends that the number of NRSA trainees supported through the MSTP program be increased from 822 in 1993 to 1,020 trainees each year by the year 1996.

Individual Fellowships

Because of the urgent need for clinical scientists familiar with patient-based research techniques, we urge NIH to increase the number of postdoctoral NRSA fellowship awards to permit the preparation of patient-based investigators.

RECOMMENDATION: The committee recommends that NIH increase support for individuals to train in patient-based research by increasing the number of

TABLE 5-5 Committee Recommendations for Predoctoral Traineeships in the Medical Scientist Training Program for FY 1994 through FY 1999.

Fiscal Year	Number of Predoctoral Trainees
1994	890
1995	955
1996	1,020
1997	1,020
1998	1,020
1999	1,020

TABLE 5-6 Aggregated Numbers of NRSA Supported Trainees and Fellows in Clinical Sciences for FY 1991 through FY 1993

Fiscal Year	Level of Training	TOTAL	Type of Support	
			Traineeship	Fellowship
1991	Number of awards	2,894	2,814	80
	Predoctoral	755	736	19
	Postdoctoral	2,139	2,078	61
1992	Number of awards	2,970	2,887	83
	Predoctoral	819	800	19
	Postdoctoral	2,151	2,087	64
1993	Number of awards	2,974	2,877	97
	Predoctoral	855	826	29
	Postdoctoral	2,119	2,051	68

NOTE: Based on estimates provided by the National Institutes of Health. See Summary Table 1.

postdoctoral fellowships in the clinical sciences from 68 in fiscal 1993 to 160 by fiscal 1996.

Institutional Training Grants in the Clinical Sciences

To permit the expansion of the pool of MSTP trainees and postdoctoral clinical science fellows, we believe modest reductions should be made in the number of postdoctoral awards made through institutional training grants. NIH reports that 2,087 awardees were supported in fiscal 1992 through this mechanism (Table 5-6). We believe a gradual decrease should occur in the number of awards (Table 5-7). This would be done to permit the expansion of the MSTP program (described above).

RECOMMENDATION: The committee recommends that the number of postdoctoral institutional traineeships supported through the NRSA program in the clinical sciences be slightly decreased from 2,051 to 1,965 between 1993 and 1996.

NOTES

1. Several studies, it must be added, have identified a lack of rigorously trained individuals who know how to perform patient-based research (e.g., Ahrens, 1992) as a special need at this time.

2. The clinical sciences are understood to include individuals holding degrees in a variety of health professions including: medicine, veterinary sciences, dentistry, nursing, clinical psychology, and social work. The research training needs of clinical psychologists have been addressed in chapter 4 of this report ("Behavioral Sciences"), dentistry needs are separately addressed in chapter 6 ("Oral Health Research"), and nursing addressed in chapter 7 ("Nursing Research"). The committee did not address research training needs in the veterinary sciences or social work, but recognizes that these fields contribute to the national research effort and merit support through the NRSA program.

3. A recent report of the Institute of Medicine, *Careers in Research: Obstacles and Opportunities* (1994) investigates ways to improve the quality of training for clinical investigators and delineates pathways for individuals pursuing careers in clinical investigation in nursing, dentistry, medicine and other health professions engaged in human research.

REFERENCES

Ahrens, E. H., Jr.
   1992 *The Crisis in Clinical Research: Overcoming Institutional Obstacles.* New York: Oxford University Press.

Association of American Medical Colleges (AAMC)
   1992 *U.S. Medical School Faculty: 1992.* Washington, D.C.: Association of American Medical Colleges.

Bradford, W.D., S. Pizzo, and A.C. Christakos
   1986 Careers and professional activities of graduates of a Medical Scientist Training Program. *Journal of Medical Education.* 61:915-918.

Fredrickson, D.S.
   1993 Clinical Investigation. Paper prepared for the Committee on National Needs for Biomedical and Research Personnel.

Frieden, C. and B.J. Fox
   1991 Career choices of graduates from Washington University's Medical Scientist Training Program. *Academic Medicine.* 66:162-164.

Institute of Medicine (IOM)
   1994 *Careers in Clinical Research: Obstacles and Opportunities.* Washington, D.C.: National Academy Press.

McClellan, D.A. and P. Talalay
   1992 M.D.-Ph.D. training at the Johns Hopkins University School of Medicine, 1962-1991. *Academic Medicine.* 67(1):36-41.

National Research Council (NRC)
   1975 *Personnel Needs and Training for Biomedical and Behavioral Research.* Washington, D.C.: National Academy Press.
   1978 *Personnel Needs and Training for Biomedical and Behavioral Research.* Washington, D.C.: National Academy Press.
   1985 *Personnel Needs and Training for Biomedical and Behavioral Research.* Washington, D.C.: National Academy Press.
   1989 *Biomedical and Behavioral Research Scientists: Their Training and Supply, Volume I: Findings.* Washington, D.C.: National Academy Press.

TABLE 5-7 Committee Recommendations for Relative Distribution of Predoctoral and Postdoctoral Traineeship and Fellowship Awards in Clinical Sciences for FY 1994 through FY 1999

Fiscal Year	Level of Training	TOTAL	Type of Support	
			Traineeship	Fellowship
1994	Recommended number of awards	2,975	2,875	100
	Predoctoral	895	875	20
	Postdoctoral	2,080	2,000	80
1995	Recommended number of awards	2,910	2,780	130
	Predoctoral	895	875	20
	Postdoctoral	2,015	1,905	110
1996	Recommended number of awards	2,860	2,680	180
	Predoctoral	895	875	20
	Postdoctoral	1,965	1,805	160
1997	Recommended number of awards	2,860	2,680	180
	Predoctoral	895	875	20
	Postdoctoral	1,965	1,805	160
1998	Recommended number of awards	2,860	2,680	180
	Predoctoral	895	875	20
	Postdoctoral	1,965	1,805	160
1999	Recommended number of awards	2,860	2,680	180
	Predoctoral	895	875	20
	Postdoctoral	1,965	1,805	160

CHAPTER SIX

ORAL HEALTH RESEARCH PERSONNEL[1]

Oral health research (OHR) has paid great dividends (Lathrop and Ranney, 1993). It has contributed to reduced rates of dental caries as well as improvements in diagnosis, prevention, and treatment of other oral diseases and abnormalities. OHR activities include research not only on tooth structure and diseases of supporting tissues, but on cells, tissues, and structures of the entire oral and craniofacial region. Other important research areas include oral cancer, salivary gland disorders, genetic diseases, and materials science related to dental practice.

Estimates indicate that dental school research training programs have less than half the faculty needed to provide a strong research environment. Compared with all other faculty in dental schools, research faculty are aging and are not being replaced at the rate they are retiring or otherwise leaving the field or by faculty having the same level of training.

Unlike other medical fields that have access to a variety of sources of research support, most OHR is supported by the National Institute of Dental Research (NIDR). Because of the anticipated demand for OHR, the number of training awards in OHR and the resultant inability to replace an aging cadre of OHR scientists in faculties of dental schools, the committee recommends a two-fold increase over the 1993 number of awards.

ADVANCES IN ORAL HEALTH RESEARCH

The history of OHR is closely related to the history of NIDR, which celebrated its 45th anniversary in 1993. Originally, NIDR's mission was to "improve the oral health of the American people." At that time, oral health focused on caries (tooth decay) because of its overwhelming prevalence. Initially, a group of intramural dental scientists took on caries research largely through epidemiologic approaches. This was the beginning of the remarkable tale of fluoride as an effective public health measure. The initial phase of epidemiology and prevention research has had a tremendous impact on every aspect of dental education and dental practice and saves the American people an estimated $4 billion per year.

In recent years the scope of OHR has been greatly broadened. In addition to the emphasis on dental caries, microbiologists and biochemists now deal with basic issues related to hard and soft oral tissues. The maturing of OHR has led not only to the understanding of how tooth decay begins, but to the improved understanding of the infectious base of periodontal diseases. These complementary discoveries have had major implications for the clinical management of oral diseases.

In addition, researchers have made much progress in areas such as neurobiology, developmental biology, cellular and molecular biology, oral microbiology and immunology, and materials science and imaging technology. Dental implants illustrate one practical application of the results of recent oral health research.

ASSESSMENT OF THE CURRENT MARKET FOR ORAL HEALTH RESEARCH PERSONNEL

A database on dental educators compiled by the American Association of Dental Schools (AADS) provides the best available current information on the OHR labor force. Begun in 1981 this database includes all faculty appointments to dental education institutions in the United States and is updated annually by a survey of AADS member institutions. It includes information on age, gender, race, academic rank, appointment status (full or part-time), academic degrees held, and area of primary appointment. Analysis of this database (Solomon, 1993) showed that the average age of OHR scientists increased from 47.3 to 49.1 years from

1986 to 1992. Also, compared with all full-time dental school faculty, OHR scientists were somewhat older (48.6 to 49.1 years, 1992-1993). OHR scientists generally held higher academic rank than full-time dental school faculty. In 1992-1993, 73 percent of OHR scientists held senior rank (full or associate professor) compared with 65 percent of all full-time dental faculty. From 1986-1987 to 1992-1993, the percentage of OHR scientists who identified their primary appointments as clinical sciences decreased from 44 percent to 36 percent (comparative figure for all full-time faculty is 58 percent). As a proportion of all OHR scientists, those holding the Ph.D. only (without a clinical degree) decreased somewhat (51 percent to 45 percent). No change was found for the dual-degree (D.D.S./Ph.D.) OHR scientists, whose proportion remained stable, but the category "other" (predominantly clinical degree with or without Masters degree) increased from 4 to 10 percent of OHR scientists (Figure 6-1).

Over the time of analysis, OHR scientists turned over (entered and left) by more than one-third (361 entered, 357 left). Most entries were in their 30s and early 40s; departures were evenly spread over the age range of the faculty. Departing faculty OHR scientists were more likely to hold a nonclinical doctorate than were entering OHR scientists.

These analyses indicate that OHR scientists are an aging group, approaching retirement. Although total numbers of OHR scientists are staying constant despite declines in total faculty, reflecting increased emphasis on research in the schools, those with Ph.D.s who leave seem to be replaced to a greater extent by those without Ph.D.s. This reflects the lack of growth in training programs for OHR since 1985. It also suggests a concern for a decrease in competitive stature for grants for OHR scientists among all research workers.

Other factors also contribute to the acute shortage of OHR scientists. Because of the undersupply of research workers for dental institutions, OHR scientists tend to 1) not take postdoctoral training to the same extent as their competitors for research grants; 2) get drafted into, or otherwise move too soon into, administrative positions; and 3) have insufficient available mentoring capability existing in the institutions where they are employed. In addition, institutional support is generally minimal, so that it is difficult to find start-up funds or bridge support.

Additional factors that contribute to the shortage of OHR scientists are similar to other areas of biomedical research. These include low funding rates for grants, lower income possibilities in academic endeavors than in practice careers, and the debt of graduating dental students. Dental students' debt is the greatest of all health care professionals: it exceeds $55,000 on average and often exceeds $100,000.

OUTLOOK FOR ORAL HEALTH RESEARCH SCIENTISTS

Although much progress has been made, oral diseases remain among the most prevalent diseases in the United States. More than 84 percent of children, 96 percent of adults, and 99.5 percent of those over 65 years of age in this country have experienced dental caries. Many millions of Americans have one or more periodontal diseases or other oral disease. Over 17 million have lost all of their teeth. In 1989, 164 million hours were lost from work and 52 million hours were missed from school because of dental conditions. In 1992, $38.7 billion was spent for dental services. By the year 2000 the annual cost for dental health is expected to reach $62 billion.

Cancer of the oropharyngeal region is more common than leukemia, melanoma, Hodgkin's disease or cancers of the brain, liver, bone, thyroid, stomach, ovary, or cervix. It affects primarily older Americans and causes approximately 8,000 deaths per year. The 5-year survival rate for oral cancer is 51 percent but only 31 percent for blacks.

Millions are at high risk for oral health problems because of other handicapping or medical conditions. These

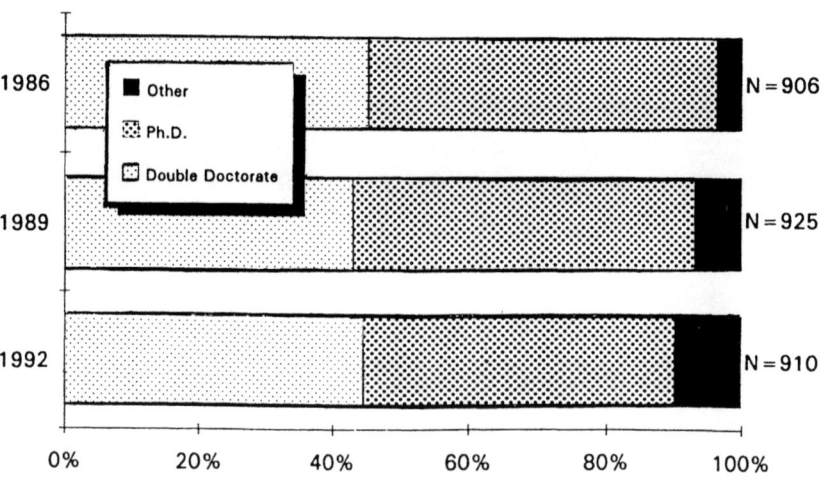

FIGURE 6-1 Percent academic degrees by academic year. SOURCE: Solomon, 1993.

conditions affect quality of life, including pain, ability to eat, speak, taste, and swallow. For example, cleft lip and palate require extensive and expensive repair to avoid disfigurement. There are significant problems for individuals with compromised immune systems, including those with acquired immune deficiency syndrome (AIDS). These problems include oral candidiasis, hairy leukoplakia of the tongue, recurrent ulcers caused by Herpes simplex or other viruses, oral Kaposi's sarcomas, and aggressive periodontal disease, including necrosis of alveolar bone. Infectious oral diseases increase risk for endocarditis, brain abscesses, pneumonia, infection of prosthetic valves and joints, and systemic infection of individuals who are undergoing organ or marrow transplants. Thus, OHR now addresses oral and dental health concerns across the life span, and high risk and special populations are major targets of this research.

OHR needs more and better-trained scientists with specific interests in oral health. Training sites and mentors must become magnets for the best and brightest of the graduates. They must also provide role models for students. Trainees must be kept in the system, and factors that facilitate retention of oral health scientists already in the work force need to be identified. It is extremely important to ensure that there are an adequate number of clinical research investigators and to enhance the clinical faculty in dentistry. Finally, research results must be disseminated to practitioners and to the public.

No funding source other than the NIDR significantly supports OHR or training specifically for OHR. Continued support of training through NIDR is essential to continued improvement of the quantity and quality of OHR personnel.

ENSURING DIVERSITY OF HUMAN RESOURCES

The percentage of women in the OHR scientist cohorts increased from 12.7 percent in 1986-1987 to 16.3 percent in 1992-1993, but the percent of women among all full-time dental faculty was still higher (18.7 percent, 1992-1993). Asians and Hispanics increased in proportion slightly among OHR scientists over the period studied, but the percentage of black OHR scientists decreased (2.6 percent to 1.7 percent). The OHR scientist group had a slightly higher representation of whites and Asians than did full-time faculty in general.

Although the OHR scientist labor force is becoming more diverse (more women and Hispanics), women and minorities are still under-represented. This is especially true for blacks.

NATIONAL RESEARCH SERVICE AWARD PROGRAM FOR ORAL HEALTH RESEARCH

Before the National Research Service Award program, dental training centered around clinical specialty training. Under the NRSA mechanism, the emphasis shifted to a primary focus on research.[2] There are now 288 people in NIDR-supported training programs for a research career, including approximately 160 in the NRSA program, excluding dental students supported under the T35 mechanism for short-term training. That number has been relatively stable throughout the 1980s and 1990s, although there has been a slight reduction in the NRSA program. Of the 288 total, approximately half are clinical researchers with a D.D.S. degree who are studying for a Ph.D. (counted as postdoctoral trainees); about 10 percent are predoctoral students studying for the Ph.D.; and the remaining 40 percent are postdoctoral trainees who hold the Ph.D. or D.D.S., including those in the NRSA program as well as in other programs (Dentist-Scientists or Physician Scientists for Dentists awards) (Table 6-1).

The NRSA stipend varies with the number of years of experience after the last professional degree and it ranges from $18,600 to $32,000. However, the Dentist Scientist Award (DSA) salary may go up to $50,000. The NRSA mechanism has a modest institutional allowance, approximately $2,500 for a postdoc on a training grant and $3,000 for a fellowship, whereas the DSA includes significant support for research of up to $75,000 over the 5-year project period and mentors can receive salary support. Under the NRSA no salary is available for mentors. Within the NRSA awards there is a distinction between the fellowships and training grants. Full tuition and fees are supplied in training grants but not in fellowships.

Legislation that limits an NRSA trainee to only 3 years of postdoctoral research experience is a significant problem for support of OHR scientists. A dentist appointed to a training grant seeking a Ph.D. needs more than 3 years, especially when concomitant clinical training is involved. Although the NIDR has been liberal in granting waivers and extensions, it would ease recruitment if the legislated limit were removed.

A good evaluation mechanism for the effects of different types of training support is needed, so that differential impacts of different programs can be known.

Programs in which dentists receive research training including the Ph.D. have been successful. Eighty-one percent of the individual and 73 percent of the institutional DSA-eligible awardees who have completed all requirements of the DSA program have obtained a placement in a health science setting (T. Valega, 1993).

Of the NRSA graduates who began their training between 1975 and 1985, between 50 and 60 percent of those who submitted project applications to NIH were subsequently funded. Clearly, both NIDR-supported NRSA and DSA trainees are able to compete successfully for grants. In Figures 6-2 and 6-3 the positive relationship between months of training and success in obtaining grants is clear. There is also a positive relationship between degree and research grant history (Figure 6-4).

TABLE 6-1 Aggregated Numbers of NRSA Supported Trainees and Fellows in Oral Health Research for FY 1991 through FY 1993

Fiscal Year	Level of Training	TOTAL	Type of Support	
			Traineeship	Fellowship
1991	Number of awards	218	186	32
	Predoctoral	78	78	0
	Postdoctoral	140	108	32
1992	Number of awards	213	178	35
	Predoctoral	77	77	0
	Postdoctoral	136	101	35
1993	Number of awards	127	105	22
	Postdoctoral	97	96	1
	Predoctoral	127	105	22

NOTE: Based on estimates provided by the National Institutes of Health. See Summary Table 1.

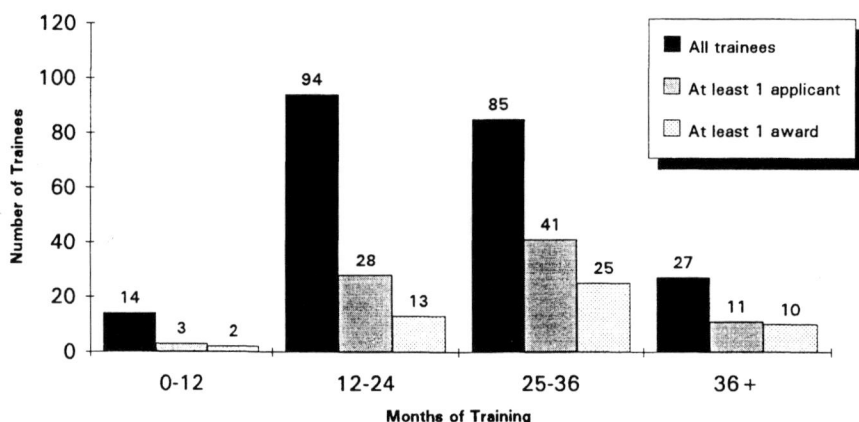

FIGURE 6-2 Individual NRSA (F32) trainees: relation between months of training and research grant history, 1980-1990 (based on individuals who began training between 1975 and 1985). SOURCE: National Institute of Dental Research, 1991.

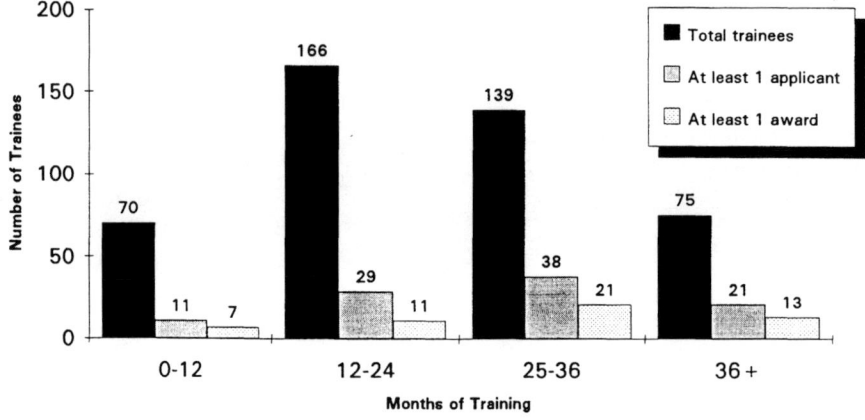

FIGURE 6-3 Institutional NRSA (T32) trainees: relation between months of training and research grant history, 1980-1990 (based on individuals who began training between 1975 and 1985). SOURCE: National Institute of Dental Research, 1991.

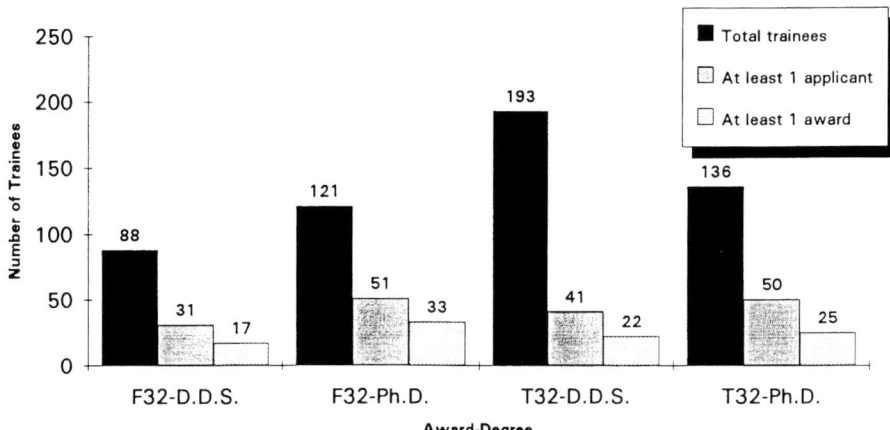

FIGURE 6-4 NRSA trainees: relation between degree and research grant history, 1980-1990 (based on individuals who began training between 1975 and 1985). SOURCE: National Institute of Dental Research, 1991.

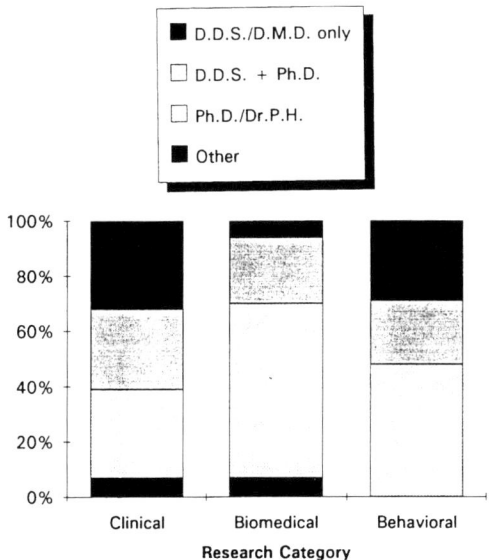

FIGURE 6-5 Percent of NIDR research grants (R01, R23, R29, R37) by degree of principal investigator and research category: FY 1990-1992. NOTE: "Other" category includes M.D., M.D. + Ph.D., D.V.M. and non-postdoctoral degrees. SOURCE: National Institute of Dental Research, 1993.

Figure 6-5 considers the types of academic training associated with the area in which research was performed. This could be used to provide an indication of the background needed for scientists to conduct clinical, basic biomedical, and behavioral research. Investigators having a dental degree, either alone or with a Ph.D., predominated for clinical research. Among biomedical research projects the investigators having only the Ph.D. predominated. No differences were found between the investigators having dental degree or only the Ph.D. for projects classified as primarily behavioral. These findings support the need for appropriately trained individuals having at least a dental degree to provide a future adequate cadre of clinical oral health researchers.

RECOMMENDATIONS

Program Size

Because of rapid advances in biological and physical sciences, the opportunity for advancing oral health has never been greater. However, there is an alarming shortage of trained researchers in oral health to take advantage of those opportunities.

Graduates of NRSA and DSA programs supply, on average, fewer than one clinical scholar or potential clinical scholar per dental school per year. At least 200 graduates per year are necessary to supply the institutions' needs (Kennedy, 1990), roughly four times the number being produced. Thus there is a shortage from two points of view: one, to address the research needs and two, to fill faculty slots with capable researchers. At least half of a dental school faculty should be clinical scholars. The other half, though perhaps not researchers, should be scholarly clinicians, that is, critical about the literature, critical about what they are teaching, and critical about their patients' needs. Research training fulfills both needs of a clinical faculty.

Looked at in another context, dental institutions are a small but important consumer of Ph.D.s who do not hold clinical degrees. However, nobody but the OHR community and dental institutions has concern for the D.D.S./Ph.D. supply that is critical to maintaining the quality and quantity of research related to oral health. Dental schools today are not able to find sufficient numbers of D.D.S./Ph.D.s to fill available faculty positions.

There are clear indications that the clinical degree is important to site of employment after training has been completed. In 1985 Littleton and his colleagues reported that 62 percent of D.D.S./Ph.D.s trained with NIDR support could be found on faculties of dental institutions, whereas only 31 percent of all NIDR postdoctoral trainees were employed in dental institutions (Littleton, 1985). It is thus important to the future of OHR that people with clinical degrees be encouraged, through adequate training support, to enter the research work force.

There is an acute need for clinical dental researchers and OHR workers in general. The National Research Council's 1985 report called for 320-400 new clinical dental research trainees annually, but the NIDR has been unable to carry out this recommendation because funds were not available. A significant increment in training would substantially alleviate the shortage of OHR personnel.

We need to produce about 260 graduates per year. This estimate includes needs for researchers in dental schools and other settings. About 30 percent of NIDR's total research support goes to non-dental school institutions. If the current distribution between NRSA and other mechanisms remains constant, roughly half should be supported by the NRSA mechanism (130 per year). At 3 - 5 years per finished trainee, this would be 390 - 650 in training through the NRSA mechanism; the current total through the NRSA is 213 and for all NIDR mechanisms is only about 300.

There is need and rationale, therefore, for a tripling to quadrupling of training for OHR (Table 6-2). Realistically, however, the need is better met incrementally rather than abruptly to ensure that existing high-quality training sites are not overloaded and to stimulate identification and development of additional high-quality sites.

RECOMMENDATION: The committee recommends that the total number of training positions available for preparation in oral health research double from approximately 210 positions in fiscal 1993 to 430 positions in fiscal 1996 and remain steady thereafter.

Need for a Dental Scientist Training Program

The Medical Scientist Training Program (MSTP) offers an integrated program of medical and graduate training leading to the combined M.D. and Ph.D. degrees. The success of that program (see Chapter 5), coupled with the demonstrated success of D.D.S./Ph.D. or D.M.D./Ph.D., suggests that OHR would benefit from the development of a Dental

TABLE 6-2 Committee Recommendations for Relative Distribution of Predoctoral and Postdoctoral Traineeship and Fellowship Awards for Oral Health Research for FY 1994 through FY 1999

Fiscal Year	Level of Training	TOTAL	Type of Support	
			Traineeship	Fellowship
1994	Recommended number of awards	260	200	60
	Predoctoral	125	100	25
	Postdoctoral	135	100	35
1995	Recommended number of awards	345	230	115
	Predoctoral	210	130	80
	Postdoctoral	135	100	35
1996	Recommended number of awards	430	265	165
	Predoctoral	290	160	130
	Postdoctoral	140	105	35
1997	Recommended number of awards	430	265	165
	Predoctoral	290	160	130
	Postdoctoral	140	105	35
1998	Recommended number of awards	430	265	165
	Predoctoral	290	160	130
	Postdoctoral	140	105	35
1999	Recommended number of awards	430	265	165
	Predoctoral	290	160	130
	Postdoctoral	140	105	35

Science Training Program (DSTP) that is analogous to the MSTP under the auspices of the NRSA legislation.

RECOMMENDATION: The committee recommends that one-quarter to one-half of the new positions available for training in OHR in fiscal 1994 and beyond be used by NIDR to establish a Dental Scientist Training Program (DSTP) under the NRSA act.

Other Considerations

Additional suggestions for improving the NRSA for training OHR scientists have emerged in committee discussions. Some of these suggestions are similar to those from other fields and are included in the overall recommendations of this report. Other suggestions are unique to OHR, and are mentioned in the paragraphs that follow.

Legislated policy requires that individual awards comprise at least 15 percent of the total NRSA allocation by the funding institute, but there is no clear rationale for the policy. NIDR sometimes has difficulty meeting the 15 percent requirement for awards to individuals, and there is great demand for institutional awards. It is possible that institutions are competing for the best students through the institutional training mechanism, leaving relatively few students for the pool of individual applicants. The committee believes that the 15 percent requirement for individual awards for NIDR should be rescinded.

The long-term effectiveness of short-term exposure to research experiences needs to be evaluated. At present, the NIDR uses the T-35 mechanism to draw dental students into research careers. However, this particular mechanism is limited to 4 percent of the NRSA funds. The committee suggests that this limitation be evaluated not only for the long-term effect but also to explore whether the mechanism should be extended for other purposes, such as retraining, encouraging minorities and women to enter research tracks, and stimulating clinical research.

Completion of a Ph.D. after a dental degree generally requires more than 3 years. Also, other sources of support for continuing such studies (e.g., Howard Hughes Institute, clinical revenues) are not generally as available for dentists as they may be for physicians. The committee suggests that doctoral support be provided for 5 years and beyond for dentists making satisfactory progress toward a Ph.D. under an NRSA.

Finally, because of the disincentive for entering research training that is inherent to the heavy debt load of dental graduates (currently the highest of all health professional graduates at greater than $55,000, on average), loan forgiveness would provide an incentive. The committee believes that a loan-forgiveness incentive should be provided as a feature of NRSA programs.

In summary, what is particularly needed in OHR are appropriately trained personnel to carry out a broadened scope of research. There is an alarming personnel shortage of research-trained full-time dental faculty. Many challenges continue to arise as the twenty-first century approaches. These challenges require enhanced resources and flexibility as well as continued cooperation and collaboration among programs and institutions if they are to carry out the mandate of Congress to improve the oral health of the American people.

NOTES

1. Much of the material in this chapter is based on the views of experts who convened a one-day workshop on July 9, 1993, in Washington, DC (Lathrop and Ranney, 1993).

2. There are two major programs for training OHR scientists through NIDR support, the NRSA and the Dentist/Physician Scientist Award (DSA). NIDR's portfolio in the NRSA includes the following:

- F-32, individual postdoctoral fellowship;
- F-33, individual senior postdoctoral fellowship for senior faculty members;
- F-35, intramural training grant;
- T-32, institutional training grant; and
- T-35, short-term summer training grant.

The DSA/PSA programs include the following awards:

- K-11, individual physician-scientist award for dentists;
- K-15, individual dentist-scientist award; and
- K-16, institutional dentist scientist award.

The F awards (fellowships) are all postdoctoral awards, either post Ph.D. or post clinical doctorates. These postdocs receive the traditional postdoctoral research training and, in the case of clinical doctorates, that may include earning a Ph.D. during postdoctoral training. The T awards (training grants) can include predoctoral students. These people have bachelor's or master's degrees and usually are in a program to obtain a Ph.D. In rare cases, such as in biomaterials or epidemiology, they stop at a master's level. The T-35 short-term grants are specifically for dental students. The physician-scientist award for dentists and the dentist-scientist awards appoint only dentists. All of these DSA appointees are in a program to obtain a Ph.D.

REFERENCES

Kennedy, J.E.
   1990   Faculty Status in a Climate of Change. *Journal of Dental Education* 54(5).

Lathrop, L. and R.R. Ranney
   1993   Proceedings of Meeting on National Needs for Oral Health Research Personnel. National Academy of Sciences, Washington, D.C., July 9, 1993. Unpublished summary. September, 1993.

Littleton, P.A., L.J. Brown, and E.S. Solomon
   1985   *The Relationship Between National Institute of Dental Research (NIDR) Supported Research Training and Careers in Dental Research.* Unpublished report. March, 1985.

Solomon, E.S.
   1994   The Oral Health Research Work-Force. To be published in *Journal of Dental Education.*

Valega, T.
   1993   Report to Meeting on National Needs for Oral Health Research Personnel, National Academy of Sciences, Washington, D.C., July 9, 1993. See Proceedings of Meeting (Lathrop and Ranney, 1993).

CHAPTER SEVEN

NURSING RESEARCH PERSONNEL

Nursing research addresses critical issues of nursing practice and health care. It seeks to understand problems of living with acute and chronic illness, factors associated with health promotion and disease prevention, and the impact of nursing systems. Research on acute and chronic illness focuses on patient and family responses to illness and disability, the biological and behavioral factors that contribute to these conditions, and ways to improve or remedy them. Health promotion and disease prevention research focuses on maintaining and improving general health status among diverse populations through effective nursing and health care practice interventions. It concerns such issues as pregnancy, health behavior, adolescent health habits, dieting practices, smoking and health behavior, as well as early detection and treatment of particular diseases or conditions. Nursing systems research focuses on patient outcomes, new approaches to quality nursing services to improve patient care, studies of innovative nursing care delivery models, and issues of clinical decision-making.

In recognition of the important role nursing research plays and the potential of its future contributions to the health and welfare of the American people, the National Center for Nursing Research, established in 1986, became the 17th institute (National Institute of Nursing Research, NINR) the National Institutes of Health (NIH) on June 17, 1993. The primary mission of NINR is to provide a strong scientific base for nursing and health care practice. Medical research focuses on finding better ways to diagnose and treat illness, and nursing research complements it by focusing on people's responses to illness and treatment. For example, a medical researcher might study how to diagnose the human immunodeficiency virus (HIV) earlier or which drugs and what doses are best to combat HIV infection. A nurse researcher might study the symptoms associated with HIV, such as muscle wasting and pain. Nursing research aims to develop interventions to maintain or improve the patient's quality of life and functional status.

With the enactment of the Health Research and Health Services Amendments of 1976 (P.L. 94-278), the NRSA training authority was extended to include the research training programs administered by the Division of Nursing of the Health Research and Services Administration. As a result, the National Research Council in 1977 described changes taking place in the composition of the nursing research labor force and training needs and specified how NRSA funds might contribute to the production of scientists in this area. Nursing research as the 1977 study committee noted in its report (NRC, 1977) complements biomedical and behavioral research by focusing on health outcomes and improved patient care.

Since the inclusion of nursing research in the NRSA program in 1976, nursing research and nursing research training have undergone many organizational changes within the federal government, ultimately resulting in the appearance last year in the National Institute of Nursing Research (NINR). NINR has a strong commitment to support research training. Its National Research Service Award (NRSA) program has primarily funded predoctoral fellows and trainees. However, there is a great need to increase the support for nurse researchers at the postdoctoral level and to shift toward a balance in the distribution of funding for predoctoral and postdoctoral fellows and trainees.

ADVANCES IN NURSING RESEARCH

Nursing research may be considered to be organized into three general categories: acute and chronic illness, health promotion and disease prevention, and systems of care. Advances in each of these three overall areas of nursing research are summarized below.

Acute and Chronic Illness

Research topics on nursing care of acute illness range from sleep-wake patterns of preterm infants in the neonatal intensive care unit to the biobehavioral factors affecting recovery after a myocardial infarction. Nursing studies of chronic illness address pain assessment and management, stress, orientation and cognitive dysfunction, and adaptation to decreased functional status (Hester, et al., 1990). They also address symptom management for diverse diseases and conditions such as fatigue, dyspnea, nausea and vomiting, anxiety, cognitive impairment, and sleep disturbances, as well as symptoms of specific diseases (McCorkle et al., 1994).

There is a renewed focus in nursing research on the application of state-of-the-art biological and behavioral theory and measurement to the clinical problems patients experience with illness and medical treatment. For example, one such study is measuring the effects of self-management biofeedback therapy on the frequency of complex ventricular ectopia, events of sudden cardiac arrest, survival rate, and enhanced heart rate variability in subjects after an episode of sudden cardiac arrest (Cowan et al., 1991). Other studies are investigating factors related to breast cancer, colorectal cancer, and depression after stroke. Investigators are combining data across studies in the areas of depression and biological immune responses (Johnston et al., in press; Tax et al., in press).

Nursing studies also examine the effectiveness of nursing procedures, such as open versus closed endotracheal suctioning. They compare different modes of oxygen delivery to patients with chronic lung disease and develop accurate tests for confirming correct nasogastric tube placement at the bedside (Rudy et al., 1991). Other procedures being investigated include feeding and handling of infants in the neonatal intensive care units, wound healing of decubiti, and behavioral strategies to maintain the functional independence of cognitively impaired nursing home residents.

Nursing research holds great promise for increasing the effectiveness of patient care and improving the quality of life for patients and their families. New discoveries enhance the potential for new and effective approaches to nursing.

Health Promotion and Disease Prevention

Nursing studies of health promotion and disease prevention examine ways in which behavioral change can be effected to improve quality of life and encourage avoidance of health risks across the life span. Descriptive and correlational studies have predominated in the past, exploring such variables as mental health in postpartum mothers, neurometric measures of premenstrual syndrome, weight management, fatigue, exercise regimens, and perimenopausal phenomena (Shaver et al., 1992). Studies of intervention programs such as educational programs for school-age children are also conducted.

Most health problems are complex and are linked to behavior and lifestyle. They require multidisciplinary approaches that combine knowledge from the biological, environmental, and behavioral sciences. Nurse investigators are seeking ways to tap into the extraordinary advances in the basic biological and behavioral sciences to better understand how healthy behaviors are established and maintained. For example, nurse researchers investigate positive and negative health behaviors in childhood and early adolescence and their linkage to individual, family, social, biological, and environmental factors. They work with colleagues in the biological sciences to test interventions such as exercise and counseling.

Women's health is another area of expanding interest for nursing research, including childbearing, nutrition, exercise, normal developmental processes, and stress and adaptation to life transitions (Woods et al., 1993). The large studies on women's health issues initiated by NIH will provide additional opportunities for nurse researchers to study the biological, behavioral, and social factors that contribute to health and disease and to test interventions that will promote health and prevent disease.

Systems of Care

Nurse researchers are evaluating the clinical context in which health care is provided, the process of nursing care, and organizational factors that affect patient outcomes and quality of care delivery. For example, the cardiovascular effects of noise and light have been studied in neonatal intensive care units and coronary care units. Other studies have documented the cost effectiveness of nurse-coordinated care from the hospital into the home for high-risk populations such as very-low-birth-weight infants, women with high risk pregnancies, and frail older people (Brooten et al., 1986).

A unique and challenging area of research is the exploration of bioethical issues in health care delivery. There is a tremendous need for empirical studies of treatment decisions, especially as they relate to advanced medical technologies and access to innovative therapies.

A major challenge for nurse investigators is the identification and measurement of interventions and clinical endpoints of nursing care that are cost effective. New modes of health care delivery and financing will continue to exert pressures to develop measures of care that can monitor appropriateness and quality of care. To answer research questions related to these issues, nurse investigators often collaborate with researchers in health services research and other disciplines.

Nursing studies utilize a variety of methods and measures, integrating knowledge across disciplines in an effort to understand human behavior, health, and disease in the context of the individual, the family, and the social and cultural environment. These complex clinical questions require the talents and perspectives of many disciplines as well as the expertise of nurses in clinical practice, who ask provocative questions about the patient care they are providing.

ASSESSMENT OF THE CURRENT MARKET FOR NURSING RESEARCH PERSONNEL

Approximately 12,000 nurses with doctoral training in the sciences were employed in the U.S. work force in 1992, up from 4,108 in 1980 (Bureau of Health Professionals, 1991). Even with this increase in numbers of doctorally prepared nurses, the gap between supply and demand remains large. Nurses often begin their research careers at midlife, with an expansion of interests grounded in clinical experience. Clinical research is a strength of nursing research, with questions for basic biomedical and behavioral research most often emanating from clinical questions being asked. Research opportunities are expanding for nurses, both within academia and in hospitals and other clinical settings (Moses, 1993). The number of nurse researchers in clinical settings is growing. In 1984, 9.3 percent of doctorally prepared nurses worked in hospitals (Moses, 1986), whereas by 1988 it was estimated that 14 percent were working in hospitals (80 percent employed as faculty and 6 percent in other areas). In 1992, 18.5 percent of doctorally prepared nurses were employed in hospital settings (63.4 percent in nursing education and the remainder in other categories). Nurses with doctorates in nursing are more likely to be employed in nursing (4,141) than in nonnursing work (100). Nurses with doctorates in related fields also tend to be employed in nursing positions (5,421), but increasing numbers are finding positions outside of nursing (1,622) (Moses, 1993). The average salary for nurses with doctorates employed in universities and 4-year colleges was $54,468 in 1992 (AACN, 1992).

There is little influence of non-U.S. citizens on the scientific labor force in nursing because most non-U.S. citizens return to their countries of origin after doctoral study.

OUTLOOK FOR NURSING RESEARCH PERSONNEL

The need for skilled research personnel in nursing is expected to expand in the decades ahead. Several new initiatives are proposed to test innovative community-based strategies to provide for health promotion and disease prevention activities, especially for high-risk populations. There is new interest in the implications of cultural diversity for health and for developing effective health care interventions. Nurse researchers have developed a track record in this area of research with vulnerable populations and have the collaborative networks established to expand in this area.

A very small proportion of nurses enrolled in doctoral nursing research programs receive NRSA support. NINR supported 261 NRSA positions in 1991. This number represented 9 percent of the total number of students enrolled in nursing doctoral programs and 22 percent of those who were full time students (Bednash et al., 1993).

Many nurses in research training programs study part time because of the lack of resources and the need to work. Financial support for doctoral or postdoctoral study in nursing research is basically nonexistent from nongovernment resources. Nurses completing doctorates in nursing report the highest percent of primary support from personal resources (73 percent) compared with the biomedical sciences (22.1 percent) and behavioral sciences (59.4 percent). Nurses also report the lowest percent of support from the university (7.2 percent) compared with biomedical sciences (52.3 percent) and behavioral sciences (32.3 percent). Federal support for nurses with degrees in nursing and the biomedical or behavioral sciences was lower than federal support for nonnurses (Survey of Earned Doctorates, 1993).

More nurses with doctorates are needed as faculty for schools of nursing. Only three-fifths (59.7 percent) of nurse faculty members with full-time appointments in schools of nursing offering doctoral programs hold earned doctorates. And only one-quarter (24.7 percent) of the faculty in institutions offering the baccalaureate degree hold a doctorate (Bednash et al., 1993). Nursing still has a great distance to go before its academic faculty are fully prepared educationally.

Another concern is the age of nursing school faculty. The mean age for faculty in schools with doctoral programs is 49.2 (Bednash et al., 1993). If young nurse scientists are not trained and recruited to academic institutions, there will be few well-qualified researchers to replace the present pool of scientists as they retire in the next two decades.

Most of the projections of need for nurses with doctoral preparation have focused on predoctoral training and have not addressed the needs of nurses with a basic research preparation to build a program of research. Because there is such a tremendous demand for nurses with doctorates in schools of nursing, faculty have many other teaching responsibilities and often delay developing an active program of research. Fewer than 50 nurses were supported in 1992 by the NRSA program for postdoctoral study.

There is also a growing demand for nurses with doctorates in service settings where the focus is on clinical research, clinical therapeutics, and quality of care. The small supply of nurses with doctoral research preparation contrasts with the large demand for more researchers to generate nursing information and necessitates formulation of a plan to revitalize funding for nursing research training.

ENSURING DIVERSITY OF HUMAN RESOURCES

Minorities

Although the number of minority nurses with doctoral preparation has steadily increased since 1985, minorities represent a very small proportion of faculty members in schools of nursing. Table 7-1 displays relative distribution of U.S. nursing school faculty by racial and ethnic composition in 1992. If we are to adequately address issues of minority representation and health concerns, recruitment of minorities into doctoral programs and as faculty members is a high priority.

Age

On average, nurses completing doctoral work are a decade older than their colleagues in other disciplines. They are moving into clinical practice after obtaining a bachelor of science degree and a master's degree, and proceeding into advanced clinical practice or academic roles before entering doctoral programs. This lock-step career progression does not encourage talented young nurses to pursue an interest in research training early in their careers. Table 7-2 shows the relative distribution of basic biomedical science and nursing work force by gender and age. In 1992 the mean age of assistant professors with an earned doctorate was 46.8 years.

Gender

Despite efforts to attract men to this area, nursing research continues to be dominated by women. Between 1985 and 1991, 96.5 percent of people with nursing doctorates were women and only 3.5 percent were male.

THE NRSA PROGRAM IN NURSING RESEARCH

Research training and career development are a major component of the long-range plan of the NINR and have consistently received strong interest from Congress. The NINR supported a total of about 255 trainees in 1991 and 1992 (Table 7-3), although more trainees were supported in 1991 and 1992 than in 1993. The number remains well below the level recommended by the National Academy of Sciences of 320 trainees by 1990 (NRC, 1985). Only 16 percent of trainees funded are in postdoctoral training, a figure which has remained relatively constant since 1988.[1]

Table 7-4 provides data on the total number of trainees (FTTPs) supported by year and by individual (F31, F32, and F33) and institutional (T32) awards through 1990. The proportion of individual fellowships to traineeships plateaued at approximately 60 percent individual and 40 percent institutional awards in 1989.

As of FY 1992, there were 18 institutional awards. The number of institutional awards is expected to increase gradually as programs of research develop within the schools of nursing that are research intensive (NINR files).

The individual award (F31 and F32) has provided nurses with the flexibility to create programs of study within geographic proximity to their homes and to cross discipline boundaries, obtaining doctorates in a variety of health-related fields. Individual trainees have been encouraged to create a team of mentors, led by one primary sponsor, who can provide the specific expertise in methods, measurement, and clinical content that fits the needs of the student.

NRSA postdoctoral support increased from 14 full-time training positions (FTTPs) (8 percent of total FTTPs) in 1986 to 42 FTTPs (16 percent of total FTTPs) in 1992 (Table 7-3). The benefits of postdoctoral study are begin-

TABLE 7-1 Relative Distribution of U.S. Nursing School Faculty by Racial and Ethnic Composition, 1992 (All Faculty)

Racial and Ethnic Category	Number	Percent
White	6,959	91.8
Black	391	5.2
Hispanic	95	1.2
Native American	20	0.3
Asian	108	1.4

SOURCE: American Association of Colleges of Nursing, 1992.

TABLE 7-2 Relative Distribution[a] of Nurses with Doctorates Compared with the Basic Biomedical Science Work Force by Gender and Age, 1991

Age Group		Nurses with Doctorates[c]	Biomedical[b]	Gender	
				Men	Women
TOTAL N		11,235	91,959	70,309	21,605
25-29	%	.5	2.4	2.0	3.9
30-34		4.8	13.1	11.3	18.7
35-39		10.0	22.2	20.7	26.8
40-44		26.9	19.9	19.4	21.4
45-49		16.0	18.4	19.4	15.1
50-54		18.6	11.6	12.9	7.4
55-59		15.0	6.5	7.5	3.3
60-64		4.7	4.0	4.5	2.3
65-69		3.6	1.8	2.0	0.9
70+		.7	0.2	0.2	0.2
Total (Median Age)		100.8 (47.0)	100.1 (42.0)	99.9 (43.6)	100.0 (39.1)

[a] In percent, may not total 100 because of rounding.
[b] Moses, 1993.
[c] Includes Ph.D. holders working in biomedical science fields and includes postdoctorates.

SOURCE: National Institute of Nursing Research, special tabulations, 1993.

ning to be recognized in the nursing research community. However, with the current funding levels of postdoctoral trainees and fellows combined with an attractive job market and the average age of 41 for nurses completing doctoral study, it is difficult to attract large numbers into the program.

To date, most of the NINR's NRSA program has supported predoctoral fellows and trainees because of the urgent need to increase the number of adequately prepared researchers. Less than 20 percent of the awards support postdoctoral research training. Efforts need to be expanded to encourage nurses who recently have completed their doctoral studies to pursue postdoctoral training. Increases in the stipends will be a first step in recognizing the need to provide financial incentives.

These issues underscore the need to incorporate creative and flexible procedures for managing NRSA awards, such as allowing interrupted appointments for other work and personal responsibilities. Multiple alternative methods for research training need to be developed to decrease the time needed for nurses to complete their doctorates after obtaining their professional nursing degrees and to enhance their research productivity.

RECOMMENDATIONS

In fiscal 1993, NINR had approximately 236 positions available for NRSA awards.[2] An increase in positions is needed to increase the number of well-qualified nurse researchers in the field and to advance the science of nursing research.

Program Size

On the basis of the number of doctoral programs in nursing and the number of research-intensive environments, an adequate infrastructure already exists within the nursing re-

TABLE 7-3 Aggregated Numbers of NRSA Supported Trainees and Fellows in Nursing Research for FY 1991 through FY 1993

Fiscal Year	Level of Training	TOTAL	Type of Support	
			Traineeship	Fellowship
1991	Number of awards	255	93	162
	Predoctoral	220	67	153
	Postdoctoral	35	26	9
1992	Number of awards	257	103	154
	Predoctoral	217	75	142
	Postdoctoral	40	28	12
1993	Number of awards	236	112	124
	Predoctoral	188	76	112
	Postdoctoral	48	36	12

NOTE: Based on estimates provided by the National Institutes of Health. See Summary Table 1.

search community to respond to an increase in the number of awards. The growth in the number of schools with research-intensive environments is expected to continue, thus providing a sufficient number of programs and resources for an expanded NRSA program in nursing research.

With the proposed changes in health care reform, continued development of a strong scientific base from nursing research for practice is essential for preparing advanced practice specialists to care for the rapidly changing needs of high-risk and underserved patient populations.

RECOMMENDATION: The committee recommends that the number of positions available for preparation in nursing research increase from approximately 236 awards in fiscal 1993 to 500 positions in fiscal 1996. These positions should be phased in yearly as properly qualified candidates and training sites present themselves (Table 7-5).

Balance in Mechanisms of Support

Because nursing research is a burgeoning field of science, there is a critical need to have an increased number of highly trained nurse researchers to develop information that is at the cutting edge for nursing practice and health care. Support for research training at the predoctoral level must be expanded to increase the number of nurse researchers prepared at the postdoctoral level. As the number of NRSA positions increase by the year 1999, there will be a progressive shift toward an eventual balance between the proportion of funding to predoctoral and postdoctoral fellows and trainees.

NRSA awards at both pre- and postdoctoral levels remain the primary mechanism for financial support for training of nurse researchers. There is a critical need to prepare an adequate pool of qualified nurse researchers to fill the

TABLE 7-4 Numbers of NRSA Supported Trainees and Fellows Supported by NINR in FY 1986 through FY 1990.

	Type of Support			
Fiscal Year	Training Grants			TOTAL POSITIONS
	Number	Positions	Fellowships	
1986	2	14	151	165
1987	9	36	131	167
1988	11	55	131	186
1989	12	69	150	219
1990	16	89	168	257

SOURCE: National Institute of Nursing Research, Special Tabulations, 1993.

TABLE 7-5 Committee Recommendations for Relative Distribution of Predoctoral and Postdoctoral Traineeship and Fellowship Awards for Nursing Research for FY 1994 through FY 1999

Fiscal Year	Level of Training	TOTAL	Type of Support	
			Traineeship	Fellowship
1994	Recommended number of awards	340	130	210
	Predoctoral	290	95	195
	Postdoctoral	50	35	15
1995	Recommended number of awards	420	160	260
	Predoctoral	360	120	240
	Postdoctoral	60	40	20
1996	Recommended number of awards	500	195	305
	Predoctoral	430	145	285
	Postdoctoral	70	50	20
1997	Recommended number of awards	500	195	305
	Predoctoral	430	145	285
	Postdoctoral	70	50	20
1998	Recommended number of awards	500	195	305
	Predoctoral	430	145	285
	Postdoctoral	70	50	20
1999	Recommended number of awards	500	195	305
	Predoctoral	430	145	285
	Postdoctoral	70	50	20

demands for faculty positions in research-intensive environments. Flexible and creative mechanisms must be established to meet the increasing demand for qualified faculty in research-intensive environments.

RECOMMENDATION: As the number of awards in nursing research expands, emphasis should be given equally to the development of predoctoral and postdoctoral awards.

NOTES

1. The predoctoral/postdoctoral split in nursing research was 86 percent/14 percent in 1988, 85 percent/15 percent in 1989 and 84 percent/16 percent 1990.

2. According to the National Institute for Nursing Research, the number of positions in 1993 was reduced due to an increase in predoctoral stipends in the absence of budget adjustments.

REFERENCES

American Association of Colleges of Nursing (AACN)
    1992   May/June 1992. Syllabus. Washington, DC.

Bednash, G., L.E. Berlin, and O. Alsheimer
    1993   1992-1993 Enrollment and Graduation in baccalaureate and graduate programs in Nursing. Washington. *American Association of Colleges of Nursing*.

Brooten, D., S. Kumar, L. Brown, P. Butts, S. Finkler, S. Bakewell-Sachs, A. Gibbons, and M. Delivoria-Papadapoulos
    1986   A randomized clinical trial of early discharge and home followup of very low birthweight infants. *The New England Journal of Medicine* 315:934-9.

Bureau of Health Professions, Division of Nursing, U.S. Department of Health and Human Services.
    1991   *National Sample Surveys of Registered Nurses, 1977-1988*. Health Personnel in the United States: 8th Report to Congress.

Cowan, J., H. Kogan, R. Burr, S. Henderson, and L. Buchanan
    1991   Power spectral analysis heart rate variability after biofeedback training. *Journal of Electrocardiology*, 23(Suppl):85-94.

Hester, N., R. Foster, and K. Kristensen
    1990   Measurement of pain in children: Generalizability and validity of the pain ladder and the poker chip tool. In *Advances in Pain Research and Therapy*, ed. D. Tyler and E. Krane, pp. 79-84. New York: Raven Press.

Johnston, C., A. Tax, A. Barsevick, F. Whitney, L. Luborsky, B. Lowery, and R. McCorkle
In press Exploration of depressive symptomatology in colorectal cancer patients, stroke patients, and psychiatric outpatients. *Psychological Review*.

McCorkle, R., C. Jepson, D. Malone, E. Lusk, L. Braitman, K. Buhler-Wilkerson, and J. Daly
1994 The impact of post-hospital home care on patients with cancer. *Research in Nursing and Health* 17:4.

Moses, E.B.
1986 The registered nurse population. Findings from the National Sample Survey of Registered Nurses, November 1984. *USDHHS, Bureau of Health Professionals, Division of Nursing*, #32.

Moses, E.B.
1993 The registered nurse population. Findings from the National Sample Survey of Registered Nurses, November 1992. *USDHHS, Bureau of Health Professionals, Division of Nursing*, Prepublication.

Rudy, E., B. Turner, M. Baun, R. Stone, and J. Bruccia
1991 Endotracheal suctioning in adults with head injury. *Heart & Lung* 20(6):667-674.

Shaver, J., M. Lentz, M. Heitkemper, and N. Woods
1992 Gastrointestinal, musculoskeletal and fatigue symptoms: Sleep in midlife women. *Sleep Research* 21:339.

Survey of Earned Doctorates
Sponsored by five Federal Agencies (National Science Foundation (NSF), National Institutes of Health (NIH), U.S. Department of Education (USED), National Endowment for the Humanities (NEH), and the U.S. Department of Agriculture (USDA), and conducted by the National Research Council (NRC).

Tax, A., A. Orsi, M. Lafferty, A. Barsevick, L. Luborsky, M. Prystowsky, D. Nahess, B. Lowery, and R. McCorkle
1994 A descriptive study of immune status in colorectal surgical patients: Lymphocyte phenotypes. *Oncology Nursing Forum*, 21(9).

Woods, N., Haberman, M., and Packard, N.
1993 Demands of illness: Individual, dyadic and family functioning. *Western Journal of Nursing Research* 15(1):10-25.

CHAPTER EIGHT

HEALTH SERVICES RESEARCH PERSONNEL

Health services research is an interdisciplinary field of research that seeks to understand the impact of organizational characteristics, financing, health personnel, and technology on the use of health services, quality of care, patient outcomes, and cost. The field draws on a wide range of disciplines, including biostatistics, epidemiology, sociology, health economics, medicine, nursing, engineering, management, and psychology. Its national importance is broadly recognized by providers, administrators, employers, insurers, and state and national policymakers who are seeking solutions to problems of escalating health care costs, erosion of access to care, concerns about the quality of care, and the overall health status of Americans. This field provides the information that is being used to design health care reform proposals and will be the source of information on the impact of any future health care reforms.

In recognition of increasing need for policy-relevant information that comes from health services research, the Agency for Health Care Policy and Research (AHCPR) was established in 1989 as an organizational locus for federal leadership and funding. In addition, health services research studies are funded by institutes at the National Institutes of Health (NIH) in specific disease categories, by the Department of Veterans Affairs, by Centers for Disease Control, and by private foundations and the health industry. Levels of funding have increased dramatically over the past 5 years, escalating the demand for well-trained researchers who can work in interdisciplinary teams.

The committee recommends that training in health services research be given higher priority and an increased allocation of National Research Service Award (NRSA) positions. In 1992, AHCPR had 92 positions, or approximately 0.5 percent of all NRSA awards. The NIH Revitalization Act of 1993 increased the allocation of NRSA awards to AHCPR to 1 percent, or approximately 180 positions in 1993. It is recommended that this be increased over the period 1994-1996 to 360 positions. Initially, priority should be given to increasing postdoctoral training opportunities as predoctoral training opportunities are expanded.

ADVANCES IN HEALTH SERVICES RESEARCH

Health services research has expanded our understanding of organizational and financial factors that affect access to care, appropriateness of services, quality, cost, and patient outcomes. Methods have been developed and applied for comparing the cost and effectiveness of alternative diagnostic and treatment technologies and for assessing the impact of health services on health status and quality of life. Advances in health services research have influenced the direction of national and state policies and have contributed to dramatic changes in the health care industry over the past decade. The rate of change in health care is ever increasing and the demand for new information on the impact of policy options and the effects of past changes is growing. The current national discussions regarding health reform are sharpening policymakers' understandings of the need for accelerating the investment in health services research and in its dissemination.

Organization and Financing

Health services research provides information on the quality and cost of alternative types of health care organizations. For example, health services research provides policy-relevant information on health maintenance organizations (HMOs) and other managed-care arrangements. Research has shown HMOs to be effective in controlling costs and providing high-quality care. However, research has also shown the rate of inflation in health care costs in HMOs to be similar to more traditional fee-for-service care. This has stimulated innovations in managed care, some of which come directly from products of research, including the use of appropriateness criteria for making coverage decisions, advances in the design of management-information systems and analytic methods for monitoring quality indicators and cost, and methods for profiling providers to compare provider performance in terms of cost and quality. These methods are supporting a new generation of managed care that emphasizes the provision of effective high-quality care at a reasonable cost.

Our knowledge of the impact of financial incentives on patient access to care, provider practice patterns, and organizational productivity come from health services research. For example, research has demonstrated the power of financial incentives and disincentives on patient care seeking and provider practice. Health services researchers developed the payment classification systems used by Medicare to pay hospitals (DRGs) and to pay physicians (RBRVs). Current research work is improving case mix and severity measurement in ambulatory care. These measures can be used to adjust payment to HMOs and other providers and represent long-term investments in health services research.

Our understanding of public health issues regarding access to appropriate and needed services comes primarily from health services research. For example, information on the growing numbers of uninsured people and their characteristics has been critical to the formulation of health care reform proposals. Furthermore, studies have shown that insurance coverage is necessary, but frequently not sufficient to ensure appropriate access to care. High-risk populations will likely require special outreach services and health education to gain the full benefit of available services. These findings are beginning to clarify and redefine the future role of public health agencies under health care reform.

Medical Effectiveness Research

New initiatives to allow better understanding of the effectiveness of health care services are advancing knowledge regarding what works, for whom, and under what circumstances. The AHCPR is supporting a range of studies on specific conditions and procedures to encourage better understanding of variations in patterns of provider practice and their consequences for patient outcomes, both clinical and patient-reported. Validated measurement scales for patient-reported outcomes, including health status and satisfaction, are products of years of research. Studies of acute myocardial infarction, cataract surgery, low birth weight, coronary artery disease, joint-replacement surgery, and other common conditions are providing new insights into the effectiveness and efficiency of our current health care delivery system as measured by improvements in patient outcomes and cost. Among the products of this research are best-practice guidelines for specific conditions that are being widely disseminated to providers, insurers, and consumers to improve knowledge and state-of-the-art practice.

Quality of Care and Patient Outcomes

Research is contributing to new and improved methods for measuring quality of health services, including their impact on patient functioning, satisfaction, and quality of life. Quality of care is broadly conceived to include relevant characteristics of the organizational structure of the health care provider, the content of the care, and the outcomes experienced by individuals with specific health problems. One of the newer areas for research pertains to the use of patient-reported measures of outcomes in conjunction with clinical measures of outcome. Patient-reported outcomes information is being applied in organized efforts to improve the total quality of services, as well as being used by some regulatory authorities (e.g., the Food and Drug Administration) for assessing quality of life effects of new treatments. The conceptual framework for measuring quality-of-care, the measures being applied, and the integration of quality measurement in organized efforts to improve quality are based on what is being learned through health services research.

Ethical and Legal Issues

The advances in medical technology and the need to provide patients with state-of-the-art care have increased legal and ethical concerns. Research is clarifying the nature of ethical concerns with new technologies (e.g., genetic screening) and providing new understandings regarding the meaning of informed consent and effective procedures for obtaining informed consent.

Among legal issues of greatest concern to providers has been the rapid escalation of malpractice claims and the overall cost of malpractice insurance. Research has shown the importance of effective provider-patient communication in reducing the occurrence of malpractice claims and has provided information that has helped shape malpractice reform legislation in several states. In conclusion, our understanding of the operation of the American health care system, its effectiveness, and its efficiency relies largely on the products of health services research. Advances in our conceptual understanding of the complexities of health care delivery, methods of measuring quality of care, cost, and patient outcomes and the development of policy-relevant information have been highly significant over the past 25 years. Even so, there are many questions unanswered regarding how best to organize and provide health care services to ensure the highest possible health status for all Americans at an affordable cost.

ASSESSMENT OF THE CURRENT MARKET FOR HEALTH SERVICES RESEARCH PERSONNEL

The need for information coming from health services research is widely recognized and growing. The size and scope of the supply of highly trained researchers are not well documented but are inadequate in the view of the committee to meet the current or projected future needs. In this section, the available information on supply is presented. This is followed by a discussion of factors contributing to an expanding need for well-trained health services researchers.

Supply of Health Services Researchers

Health services research is problem-oriented: practitioners examine and evaluate the delivery of health care services in the United States. It is a distinct area of inquiry in which systematic methods are applied to problems of the allocation of finite health resources and the improvement of personal health care services. Individuals enter the field of health services research from a variety of backgrounds, including biostatistics, epidemiology and bioengineering, the behavioral sciences (anthropology, sociology, and psychology), the social sciences (economics, statistics, and urban planning), and other fields such as operations research, industrial engineering, public administration, health education, and medicine.

The number of health services research personnel in the U.S. labor force is not known, although attempts have been made by the National Research Council (NRC) and others to describe the composition of certain segments of the labor force.[1] The composition of health services research is largely determined by the availability of support for research and development. Thus, no stable estimates of the number of individuals in the work force can be generated unless specialized surveys are conducted or special estimates are generated, as demonstrated by previous NRC surveys. Perhaps the best available estimate of the size of the health services research labor force is the membership of the Association for Health Services Research (AHSR), an organization established in 1981 to promote the field of health services research (Davidson, 1993). Although this most likely results in an undercount of the health services research labor force, studying the number of AHSR members and their characteristics is helpful in understanding the infrastructure of at least one segment of the labor force, a segment we might consider to be the "attentive" workers.[2]

Composition of the Attentive Health Services Research Labor Force: 1992

About 2,000 individuals were members of AHSR in 1992. Of these, 100 were students and the remainder were drawn from medicine, public health, and research backgrounds. Most AHSR members hold doctoral degrees, whether research doctorates (922 members) or clinical doctorates (448 members) (Table 8-1).

When asked to identify their discipline of specialization,

TABLE 8-1 Distribution of Degrees Among Members of the Association for Health Services Research: 1992

Degree	Number
Doctorate	922
Masters	720
(Public Health)	(247)
(Other Masters)	(473)
Bachelors	78
Practitioner	601
(M.D. or D.O.)	(448)
(Nursing)	(108)
(M.S.W.)	(17)
(Other)	(28)
J.D.	24
TOTAL	2,345[a]

Source: Davidson, 1993
[a]Total exceeds 2,000 owing to multiple degrees of some members.

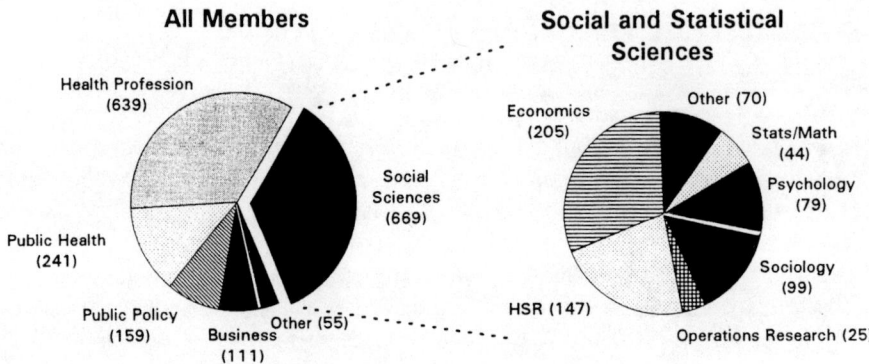

FIGURE 8-1 Disciplines of individual members (members with multiple disciplines are counted more than once), 1992. SOURCE: Association for Health Services Research, 1993.

about one-third indicated that they work in what AHSR considers to be the social sciences, another third in the health professions, and the remainder in public health, public policy, or other health services research specialties. Among those classified as social scientists, about 33 percent work in economics, 15 percent in sociology, and 12 percent in psychology. About 22 percent (147 members) of the social scientists reported their discipline to be health services research, possibly reflecting the participation of that cohort of individuals formally trained in health services research in recent decades (Figure 8-1).

Employment Sector. Most AHSR members were employed in university settings in 1992 (Figure 8-2). However, a host of non-profit and propriety health services research firms and non-profit professional organizations offer employment to health services research personnel. In 1988, the AHSR, together with the Federation for Health Services Research published the *Directory of Health Services Research Organizations.* That directory remains "the only source of information on health services research centers in the United States" (Davidson, 1993).

Changes in Composition Over Time

While the lack of data sets prevent an analysis of the composition of the health services research labor force over time, the availability of previous work by the NRC (1977, 1985) suggests that comparative studies might be developed. Because of the dynamic nature of this "labor force" and the importance to the national health effort, the Committee believes that some investment in labor force studies of the health services research community would yield tremendous payoffs—and, given the direction of national interest in the improvement of health care delivery, will be increasingly sought in the coming years.

OUTLOOK FOR HEALTH SERVICES RESEARCH PERSONNEL

The President's proposal for health care reform as well as Congressional proposals are pressing for change in the American health care system to remedy problems of spiraling costs, eroding access for the uninsured and underinsured, and uncertainty regarding the uniform quality of services. Many aspects of these proposals draw on information derived from health services research, as discussed above. Market forces are already requiring the pharmaceutical industry to consider issues of cost-effective outcomes of treatment as part of their business. However, as efforts are made to predict the consequences of alternative health care reform proposals, it is evident that much more information is needed if we are to make informed policy choices. Congress and the president have recognized this need and have increased substantially the funding of health services research by federal agencies and are expected to continue to increase funding into the future. Parallel increases in the funding of health services research in the private sector are occurring and are likely to continue.

Questions to which better answers are needed include a spectrum of issues that cut across all health care services. These include:

• Which models of health care organization and financing work best and how does this vary across populations with different socioeconomic, ethnic, and health status characteristics?

• What impact do alternative organizational models

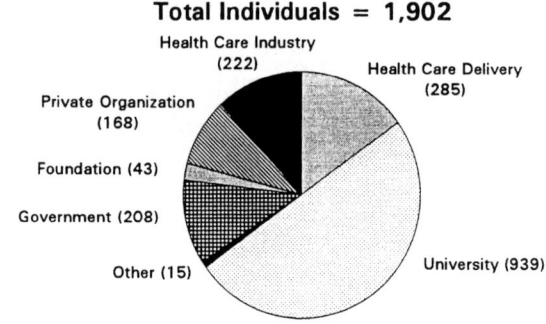

FIGURE 8-2 Employment settings of members, 1992. NOTE: Data represents 98% of AHSR membership. SOURCE: Association for Health Services Research, 1993.

have on the provision of preventive services, acute and long-term care, and quality of care and costs?

• What are the effects of licensing and regulatory mechanisms on access to care, quality, and cost and how should these rules be changed, if at all?

• Which health care professionals, with what types of training, are needed to provide high quality and efficient primary care, specialty services, long-term care, and rehabilitative services?

• How can accountability be improved? Can a useful report card on quality of care and costs be provided to patients and consumers to assist in making informed choices? What are other mechanisms to improve accountability in the health care system?

• What would be the impact of global budget constraints? Are there other means for reducing the rate of cost increases while providing appropriate services to all who need them (e.g., eliminating inappropriate services and reducing administrative inefficiencies)?

These and other questions are sharpening the issues that need to be addressed through interdisciplinary health services research and demonstration studies. As changes continue to occur in the provision of health care services, there will be growing needs for rigorous evaluations of the impact of innovations on the quality, cost, and patient outcomes of care. Evaluations studies can clarify positive and negative aspects of innovations in health care delivery as well as point to opportunities to introduce improved models of care. Numerous organizational and financing changes are being introduced by states that are seeking tailored approaches to health care reform. These state initiatives also will need information on which to shape policy and evaluate progress. The source of this information is the field of health services research.

Future Employment Conditions

The growing demand for health services researchers can be seen in multiple areas, including both public and private sectors. The AHCPR was created in 1989. Since then its budget for research has doubled and is expected to continue to grow as the demands for information relevant to health care reform increase.

Other federal agencies fund health services research, but AHCPR is viewed as the lead agency. Among the NIH institutes, there are increasing commitments to health services research. NIMH, National Institute of Alcohol Abuse and Alcoholism, and the National Institute on Drug Abuse were mandated by Congress to spend 12 percent of their budgets in 1993 and 15 percent in 1994 and 1995 to support services research in their respective areas. In 1994 this will amount to over $200 million in research funding. Other institutes fund health services research but do not identify it as such. The National Cancer Institute, National Heart, Lung, and Blood Institute, and National Institute on Aging all have significant research activities concerned with the provision of services and the effectiveness of care within their categorical disease responsibilities. It is anticipated that health services research funding will grow to represent a small but significant proportion of NIH funding. NIH currently spends substantially more on health services research than does AHCPR and is expected to continue to be a major source of funding for studies concerned with the organization of services, treatment, and outcomes of care for individuals with specific diseases and injuries.

In addition to NIH, other components of the Department of Health and Human Services fund health services research. The Centers for Disease Control are making new investments in preventive services research and the Office of Research and Demonstrations at the Health Care Financing Administration supports a substantial health services demonstration and evaluation research program. Other federal agencies also fund health services research, including the Department of Veterans Affairs, which has an expanding health services research program.

The investment of private industry in health services research also is rapidly growing. Insurers are seeking improved methods for reviewing claims and profiling providers. Managed care organizations are investing in improved methods for monitoring services, provider practices, and patient needs. The pharmaceutical industry is assessing outcomes of care related to drug therapies by using health status instruments and is investing in cost-effectiveness studies to demonstrate the comparative benefits of treatment. Every indication is that these investments will increase as managed care organizations demand better information to guide decisions regarding preferred treatments, appropriateness of services for different patients, and their impact on total costs of care.

The health care industry accounts for 14 percent of the gross domestic product and is growing, possibly reaching 19-20 percent of the GDP early in the next century. Not only is it a large domestic industry, it is a significant source of international trade. Methods and techniques developed in America (e.g., DRGs for hospital payment) are being adapted and used in many other countries. Health status measurement indices developed here are being translated and validated for use in other languages. There is general consensus that there are similar problems being faced by all nations as each attempts to meet the growing needs for heath care services. Our past investments in health services research have made us an international leader. Other countries have begun to make investments in this area because of their needs for information and the demonstrated success of the American investments.

ENSURING DIVERSITY OF HUMAN RESOURCES

There is limited information currently available on the supply of health services researchers. Efforts need to be made to improve the completeness of information on individuals being trained in health services research and actively involved in research careers. On the basis of available information, it appears that career opportunities are open for women and men in this field. However, there is no information available regarding ethnic minorities. It is recommended that efforts be made to expand opportunities for ethnic minorities to pursue education and careers in this field. The NRSA awards can be used to leverage increased diversity among individuals entering this field and this should be encouraged.

Training of Health Professionals

Another concern relates to the lack of active involvement of the full range of health professions in the conduct of health services research studies. Much of past research has focused on physician-provided services or those delegated by physicians to physician assistants or nurse practitioners. The breadth of research needs to be expanded to include services provided by the full range of health professions, including occupational therapy, optometry, podiatry, physical therapy, and social work. In addition, services provided by practitioners of alternative medicine need to be included. A recent national survey reported that one-third of all Americans have used alternative medicine services and paid for most of this care out-of-pocket.

One way to increase the diversity of research on the full range of health services provided in this country is to attract practitioners in these professions into health services research careers. The NRSA awards should be used to accomplish this goal.

THE NRSA PROGRAM IN HEALTH SERVICES RESEARCH

The information on current NRSA awards and the recommendations for future award levels in health services research relate solely to those awards made by AHCPR. The numbers of awards and funding levels of the program within AHCPR are shown in Table 8-2. The numbers of awards have grown rapidly yet remain small relative to the total needs for health services researchers. The current AHCPR program funds about 35 predoctoral and 59 postdoctoral positions, up from only 12 postdoctoral awards in fiscal 1990. The AHCPR program is relatively young and has given priority to postdoctoral training, particularly of health professionals, as an efficient strategy for more rapidly expanding the numbers of qualified health services researchers. It is expected that this should change over time to give increased emphasis to predoctoral training of individuals for careers in health services research. This training may occur either in academic disciplinary departments that have the capacity to train in health services research or in academic health services research departments that draw together faculty representing the range of disciplines applied in this field.

NRSA awards in the NIH institutes also support the training of some health services researchers. NIMH supports

TABLE 8-2 Aggregated Numbers of NRSA Supported Trainees and Fellows in Health Services Research for FY 1991 through FY 1993

Fiscal Year	Level of Training	TOTAL	Type of Support	
			Traineeship	Fellowship
1991	Number of awards	12	0	12
	Predoctoral	0	0	0
	Postdoctoral	12	0	12
1992	Number of awards	94	82	12
	Predoctoral	35	35	0
	Postdoctoral	59	47	12
1993	Number of awards	96	79	17
	Predoctoral	30	30	0
	Postdoctoral	66	49	17

NOTE: Based on estimates provided by the National Institutes of Health. See Summary Table 1.

training in mental health services research, and the National Institute of Alcohol Abuse and Alcoholism is now soliciting training grant proposals in alcohol services research. In addition, the National Institute of Nursing Research supports training in health services research and to a lesser extent so do other institutes. NRSA training supported by the institutes should be encouraged; the numbers of current trainees is unknown because the classification of training awards in behavioral, clinical, nursing, and oral health does not discriminate which programs and fellows are pursuing training in health services research methods and their application.

RECOMMENDATIONS

Program Size

Health services research is critical to the future of health care delivery in this country. Health services research is a relatively young field that uses interdisciplinary approaches to examine the impact of organization, finance, and use of technology on the utilization, cost, and quality of care. This field of research will need to grow substantially to meet the ever expanding demands for information by policymakers, administrators, providers and consumers. The questions raised regarding what impact different proposals for health care reform will have on access, cost, and quality of care are largely questions that will be answered by this field of research (Table 8-3).

RECOMMENDATION: The committee recommends that the number of NRSA positions allocated to AHCPR increase to 360 in fiscal 1996. These positions should be phased in yearly as properly qualified candidates and training sites present themselves.

Traineeships and Fellowships

The institutional training grant permits the development of innovative interdisciplinary research training programs,

TABLE 8-3 Committee Recommendations for Relative Distribution of Predoctoral and Postdoctoral Traineeship and Fellowship Awards for Health Services Research for FY 1994 through FY 1999

Fiscal Year	Level of Training	TOTAL	Type of Support	
			Traineeship	Fellowship
1994	Recommended number of awards	115	95	20
	Predoctoral	55	45	10
	Postdoctoral	60	50	10
1995	Recommended number of awards	240	145	95
	Predoctoral	180	95	85
	Postdoctoral	60	50	10
1996	Recommended number of awards	360	190	170
	Predoctoral	300	140	160
	Postdoctoral	60	50	10
1997	Recommended number of awards	300	140	160
	Predoctoral	60	50	10
	Predoctoral	60	50	10
1998	Recommended number of awards	360	190	170
	Predoctoral	300	140	160
	Postdoctoral	60	50	10
1999	Recommended number of awards	360	190	170
	Predoctoral	300	140	160
	Postdoctoral	60	50	10

an essential feature of research in this area. However, given the anticipated growing demand for skilled specialists in health services research, the committee concludes that AHCPR should place significant emphasis on individual fellowships in the next few years in order to encourage qualified individuals with some experience in the area of health care policy to pursue advanced training.

RECOMMENDATION: The committee recommends that individual fellowships represent about 45 percent of total NRSA support available through AHCPR in fiscal 1996, up from approximately 15 percent in fiscal 1993.

NOTES

1. For example, in 1977 the NRC identified and surveyed about 900 individuals who had received support from the National Center for Health Services Research (NCHSR) between 1960 and 1976 and about 1000 individuals who had received research training support from NCHSR or Alcohol, Drug Abuse, and Mental Health Administration in the area of health services research between 1970 and 1977. About 77 percent of the former trainees and 81 percent of the former principal investigators were engaged in health services research at the time of the survey. (See NRC, 1977; Ebert-Flattau, 1981).

2. The concept of the "attentive" public was developed by Gabriel Almond, who applied it to understanding attitudes of Americans toward foreign policy issues (Almond, 1950). Jon Miller applied the concept to the formulation of science policy and has expanded the original conception into a broader model of political specialization (Miller, 1983). The concept is extended for use here to refer to those members of the health services research labor force sufficiently interested in being identified as members of the field to have become members of AHSR.

REFERENCES

Association for Health Services Personnel
    1993   *AHSR Membership Directory.* Washington, D.C.: Association for Health Services Research.

Almond, G.
    1950   *The American People and Foreign Policy.* New York: Harcourt, Brace and Company.

Davidson, B.
    1993   *Personnel Needs and Training for Health Services Research.* Paper prepared for the Committee on National Needs for Biomedical and Behavioral Research Personnel.

Ebert-Flattau, P.
    1981   Some preliminary data on the health services research labor force in the United States. In *Systems Science in Health Care*, C. Tilquin (ed.), New York: Pergamon Press.

Miller, J.
    1983   *The American People and Science Policy.* New York: Pergamon Press.

National Research Council
    1977   *Personnel Needs and Training for Biomedical and Behavioral Research.* Washington, D.C.: National Academy Press.
    1985   *Personnel Needs and Training for Biomedical and Behavioral Research.* Washington, D.C.: National Academy Press.

CHAPTER NINE

RECOMMENDATIONS AND REMAINING CONSIDERATIONS

The key feature of the National Research Service Award (NRSA) continues to be its ability to influence the quality and direction of research training within the biomedical and behavioral sciences. Its ability to promote multidisciplinary training provides a multiplier effect within graduate programs. That is, the organization of the training experience within a program under the auspices of the NRSA can bring several disciplines to bear on the training of a single individual. The NRSA is also able to leverage the recruitment of minorities and women into research careers and influence the issues that will be taken up by the research community and the way in which that research will be conducted. In areas such as the behavioral sciences, where students have depended heavily on teaching assistantships to provide for graduate support, the NRSA reduces the necessity for those commitments and can thereby facilitate the completion of doctoral studies in those fields.

The committee notes, however, that there are significant weaknesses in the design of the NRSA program. For example, NRSA stipends are not competitive with other stipend sources, and training grant directors find themselves further frustrated by their inability to supplement those stipends with federal funds. These stipends are taxable, which further devalues these awards. Sensing some of these shortcomings, the U.S. Congress and the National Institutes of Health (NIH) have introduced changes in the NRSA program in recent years designed to respond to emerging education and employment challenges. In 1993 the U.S. Congress authorized an increase in the number of NRSA trainees in response to the an increase in the number of students dropping out of doctoral degree programs during the previous decade. In the same Act (P.L. 103-43), Congress revised the payback provision by restricting it to postdoctoral support with the idea that the new arrangement would encourage the participation and retention of physician-researchers. Likewise, aware of the diminishing competitiveness of NRSA stipends in attracting the most able scientists to health research, NIH recently proposed to the Department of Health and Human Services that NRSA stipends be increased at the predoctoral level to $10,000 and at the first-year postdoctoral level to $19,600.[1]

Each of these actions reflects a commitment on the part of the federal government to enhance the effectiveness of the NRSA program. The committee shares this commitment and has identified further modifications to the NRSA program to increase its effectiveness in meeting national needs for biomedical and behavioral scientists. The features of the NRSA program that merit immediate consideration are stipend support, multidisciplinary training, and flexibility in postdoctoral awards.

STIPEND ISSUES

Although interest in doing research and long-range employment prospects supply compelling reasons for pursuing a research career, more immediate incentives, such as stipends, play an indisputable role. With that in mind, we find it disturbing to note that stipend levels for predoctoral trainees in the NRSA program have remained unchanged since 1991 at $8,800 taxable salary per year.[2]

Numerous speakers at our public hearing in May 1993 voiced their concerns about the inadequacy of NRSA stipends. The existing structure of a $700 monthly stipend is simply not sufficient. Many state university stipends, for example, start at $11,000 and National Science Foundation currently pays $14,000. According to some speakers at our public hearing, most universities must work hard to supplement predoctoral stipends to raise them to $14,000 because supplementation from other federal sources (e.g. research grants) is not allowed. When universities are unable to augment stipends, trainees are forced to seek other forms of part-time employment drawing them away from their academic program and extending their time to degree.

Postdoctoral awardees do not fare much better, earning approximately $18,600 in their first year of training and $19,700 in their second. It becomes very difficult at this important period of training to entice a clinician, already burdened with debt, into a research career with a considerable reduction in compensation in order to pursue preparation as a scientist. Not only should stipend levels be increased to make them competitive, but the training budget should be sufficient to allow annual cost-of-living adjustments computed into each training grant's continuation base, with due consideration to differences in costs by region, as suggested by public hearing participants.

RECOMMENDATION: The committee recommends the NRSA stipend support at the predoctoral level be increased to $12,000 and first-year postdoctoral stipends increased to $25,000 (both adjusted for inflation) by fiscal 1997. In addition, there should be a yearly cost-of-living increase in NRSA stipends. This expansion in stipend support should be achieved through the addition of funds to the current NRSA training budget.

Estimating Program Costs

The committee has developed cost estimates for implementing these stipend increases, and balanced consideration of stipend increases against its numerical recommendations. In developing these estimates, the committee has had to make certain arbitrary decisions. First, we chose fiscal year 1993 as the base period because it is the most recent year for which reliable program data are available. Second, we developed numerical recommendations for the period 1994 through 1999 to overlap with the next assessment of the NRSA program scheduled for release in 1998.

Much of the increase in program costs which result from our recommendations is concentrated in the years 1994-1996, during which time stipend costs rise at an average annual rate of 7.8 percent per year. This results from the Committee's recommendations to increase both the number of awards and stipend levels during that period. Annual growth in program costs from 1997 through 1999 is about 2 percent each year. This reflects the committee's recommendation to keep the number of awards constant and to limit stipend increases to the expected rates of inflation. Details regarding these calculations may be found in Appendix H.

ENHANCING THE EFFECTIVENESS OF THE NRSA PROGRAM

Flexibility in Career Training at the Postdoctoral Level

In May 1993 we convened a public hearing to invite suggestions for increasing the effectiveness of the NRSA program. Most of those testifying on the role of the NRSA program in recruiting women said that the program must be more flexible in the areas of part-time training, reentry training, family leave, and geographic location of training sites. Committee members have also been concerned, however, that there is a disparity between the number of women receiving NRSA training and the number of recipients of NIH research grants. For example, women account for over 40 percent of the Ph.D.s produced in the life sciences, but they make up only 15 percent of the funded principal investigators. The problem would seem to lie not with talented young women moving into science, but rather with the development of their careers.

Women appear to be leaving science between the time they receive their doctorate and the time that they fully establish themselves in a research career track. Women are slightly less successful than men in obtaining FIRST awards, and proportionately fewer women apply for funding at certain career stages (NIH, 1993a). Therefore, there is a need to foster a supportive environment for career development. Institutional responsibilities should include mentoring, career advising, grant-writing training, and advising on transitions along the career path. The NRSA program can clearly play a role in fostering the careers of these scientists.

There is a need, then, to reshape NRSA awards at the postdoctoral level to encourage women to fully utilize their research talents. NRSA awards should allow retraining and career shifts to help women who have stopped out of research to update skills and move into emerging areas. More reentry options should be provided through NRSA programs.

RECOMMENDATION: The committee recommends that the NIH examine research training opportunities for women through the NRSA program and strengthen the role of postdoctoral support to assist women in establishing themselves in productive careers as research scientists.

Monitoring Progress Toward NRSA Goals

The NRSA is one of many sources of support available to individuals pursuing advanced preparation in research at the doctoral or the postdoctoral level. Designed to augment Federal support through the research assistantship, the NRSA program was designed to select qualified candidates from the pool of graduate students and postdoctoral personnel, provide them with a period of intense and advanced training, and launch them into productive research careers.

Perhaps one of the most significant findings of this committee is the general lack of information about the outcome of the NRSA program given almost two decades of support. Very little serious evaluation of the NRSA program has been undertaken with the support of NIH, except for a few student outcomes studies undertaken by earlier NRC committees (Appendix A). We cannot underscore strongly enough the need for follow-up information to assess program outcomes. In part, this involves the organization of existing files at NIH to permit the analysis of program outcomes. In part, the analysis that is needed will require serious review of data collection and analytic capabilities at NIH and the development of new strategies to assess career outcomes. Nowhere is the need for accurate information more evident than in our inability to track the participation of underrepresented minorities in the biomedical and behavioral research effort.

Data on Minority Participation in Science Careers

Present NIH data-collection procedures make it difficult to assess minority participation in various programs and to evaluate program effectiveness.[3] From this perspective, four data concerns emerge[4]:

1) NIH does not have a standard taxonomy for race and ethnic origin. The Public Health Service Form 398, which is used for competing research grants and for NRSA institutional training grants, specifies the following categories: American Indian or Alaska Native, Asian or Pacific Islander, black (not of Hispanic origin), Hispanic, and white (not of Hispanic origin). However, the information collected on the Statement of Appointment Form 2271 splits the Asian category into Asian (not a Pacific Islander) and Pacific Islander. This is a useful distinction because Pacific Islanders are generally considered to be underrepresented in biomedical research whereas other Asians are not. The differences in categories are puzzling.

2) The format of the tear-out page in the training grant and fellowship applications on which applicants identify their race and/or ethnic origin and gender appears to result in varied and often very low response rates. Applicants are not obligated to respond and, moreover, are told that the sheet will be separated from the application and that the information will be entered into the central NIH database and used for aggregate statistical analyses.

3) Once the racial and ethnic data are encoded in the NIH master file, it is not readily available to program officers. One cannot, for example, obtain a racial and ethnic breakdown of NRSA predoctoral fellows from the program office. The only information that program offices have comes from an institutional certification that a fellow is an underrepresented minority but does not specify which group.

4) A fourth problem common to many programs within NIH and in other agencies is ambiguity about whether the terminology should focus on minority or underrepresented minority. Although, application materials usually specify "underrepresented minority" (in targeted programs), it is not clear that the distinction is made consistently. Applicants for NRSA institutional training grants frequently mention Asians in reporting information although they are not one of specified target groups in the Minority Recruitment Plan. The distinction is crucial. Discussion in the NIH databook on minorities in extramural grant programs[5] states that minorities received over 8 percent of total research grants but underrepresented minorities received only 2.7 percent of such awards (NIH, 1993b). The committee concludes that many of these shortcomings could be addressed if all questionnaires designed to collect information about NRSA recipients use the same racial and ethnic categories.

The Need for Well-Designed Career Outcomes Studies

Improving the effectiveness of the NRSA program will require attention to issues not new to the research community. However, with the inevitable changes that will occur with health care reform and budget deficit reduction, NIH may find itself in a position of justifying its support for evaluation research. The NRSA program goes hand in hand with the government's role in financing fundamental research. This combination of research and training support has considerable and continuing benefits to the health and welfare of the citizens of our country. Well-designed career outcomes studies can provide the kind of feedback that is needed to ensure that the NRSA program is both efficient and effective given constraints being placed on the federal funding effort.

The Need for Studies of Institutional Impact

NRSA institutional training grants play an important role in providing funds to universities for student support. The committee heard testimony at its public hearing in May 1993 which suggests, however, that NRSA support may have the effect of inducing academic institutions to reduce (or not expand as much as they would otherwise) their own institutional funds devoted to the support of graduate students. Research by Ehrenberg, Rees, and Brewer (1993) suggests that this type of behavior can have a small but substantial effect on patterns of support. "Compensation" of this sort is particularly likely to happen at large research universities. Future committees would benefit from more studies of the impact of NRSA support on the recipient institution's total pattern of training support.

RECOMMENDATION: The committee recommends that the NIH review its data bases as management information systems and introduce changes in data collection, analysis, and dissemination to permit more effective tracking of NRSA award recipients. Emphasis should be given to the analysis of minority participation in research and training. New funds should be directed to the evaluation of NRSA program outcomes, including studies of the impact of institutional support on graduate student support patterns at U.S. universities.

NOTES

1. The results of that request were not available at the time of committee discussion.

2. The committee notes for comparison that the federal government set the "poverty level" in 1990 at $6,257 for a one-person household below age 65, according to the U.S. Census Bureau, 1992.

3. In 1989 the NRC study committee outlined in detail a strategy for evaluating the NRSA program, but NIH was unable to initiate any evaluation studies until 1993, when a study of the Minority Access to Research Careers (MARC) program was launched. The MARC program has served and should continue to address the important goal of strengthening the training capabilities of undergraduate minority institutions as well as training minority students who might choose research careers in the biomedical and behavioral sciences.

4. The information in this section was drawn from a commissioned paper by Sharon Bush. See Appendix D for a list of other contributors.

5. For a copy of this report, contact Marie Chang, Division of Research Grants, National Institutes of Health, 301/594-7328.

REFERENCES

Ehrenberg, R.G., D.I. Rees and D.J. Brewer
    1993   How would universitites respond to increased federal support for graduate students? In *Studies of Supply and Demand in Higher Education*, ed. C.T. Clotfelter and M. Rothschild, pp.183-206. Chicago: University of Chicago Press.

National Institutes of Health (NIH)
    1993a   *NIH Data Book: 1993*. Publication No. 93-1261. September. Bethesda, MD: National Institutes of Health.
    1993b   *Minorities in NIH Extramural Grant Programs, Fiscal Year 1982-1991*. SAES, ISB, DRG of NIH. Bethesda, MD: National Institutes of Health.

APPENDIXES

APPENDIX A

HISTORICAL OVERVIEW[1]

For nearly 20 years, the National Research Council (NRC) has played an active role in the ongoing review of training opportunities available to individuals seeking advanced preparation in the biomedical and behavioral sciences. During this time, the NRC has issued nine reports which describe the optimal structure of the National Research Service Awards Program (NRSA) given national requirements for health-related research scientists and available training opportunities.

The historical summary that follows reports information in three areas: (1) a brief history of NIH support for research training in the biomedical and behavioral sciences, including trends in support of the NRSA program; (2) an overview of the analytic contributions of previous NRC study committees; and (3) a summary of findings from studies evaluating the outcomes of the NRSA program conducted under the auspices of previous NRC committees.

LINKING HEALTH RESEARCH WITH TRAINING

The National Institutes of Health (NIH) were authorized as early as 1930 to support the training of health scientists. Specifically, Public Law 71-251, the "Ransdell Act", formally established the "National Institute of Health" as a federal agency and directed the agency to recognize the training of scientists as one of its major responsibilities. Under the terms of the Act, individual scientists could be designated to receive "fellowships" for duty at the National Institute of Health or to conduct investigations at "other localities or institutions in this and other countries" (Lenfant, 1989).

It was the enactment of the National Cancer Act of 1937 (P.L. 75-244) that established the first disease-specific institute at NIH and led to the formation of the first major program of fellowship support by the U.S. government. The National Cancer Act instructed the NIH to provide stipends or allowances to "the most brilliant and promising research fellows from the United States or abroad ... for training and instruction in technical matters relating to the diagnosis and treatment of cancer" (P.L. 75-244). Initial training efforts focused on postdoctoral research fellows and clinical training for physicians "to improve their capability in diagnosis and therapy" (Ahrens, 1992).

Federal interest and involvement in biomedical and behavioral research increased dramatically after World War II largely as a result of the demonstration during the war of the immediate and beneficial impact of well-organized basic and clinical research in meeting "national needs" (NRC, 1976). Congress concluded that human health and well-being of all Americans would benefit from the infusion of substantial sums to support research conducted by highly skilled investigators, and the Public Health Service Act of 1946 (P.L. 79-487) provided explicit authority for grants to support the training of research scientists. Thus, with the establishment of each institute, the authority was granted for the institute to train individuals in the diagnosis, prevention and treatment of disease. The fundamental assumption which links federal responsibility for research to a responsibility for training is that the quality of research depends on the talents of individuals attracted to a career in research.

Initially, the scientists needed for the health research effort were trained at the postdoctoral level, either as a result of attracting scientists from other fields to new subject areas or as a result of a need to further hone the skills of talented graduates. However, owing to the increasing demand for well-prepared research scientists, federal interest in the training of biomedical and behavioral scientists expanded in the 1950s to include the support of graduate students at the predoctoral level. "Grantee institutions" were permitted significant latitude in the management of predoctoral train-

APPENDIX A

APPENDIX TABLE A-1 Total Number of NIH and ADAMHA NRSA Research Training Positions, FY 1976-1993

Year	NIH Trainees	NIH Fellows	NIH Total	ADAMHA Trainees	ADAMHA Fellows	ADAMHA Total	Total NIH & ADAMHA	Positions Recommended by NRC
1976	8,141	1,513	9,654	N/A	N/A	1,896	11,550	13,901 [a]
1977	8,412	1,786	10,198	N/A	N/A	1,793	11,991	13,925 [a]
1978	9,360	1,863	11,223	N/A	N/A	1,709	12,932	
1979	9,204	1,993	11,197	N/A	N/A	1,533	12,730	
1980	8,878	1,786	10,664	N/A	N/A	1,393	12,057	12,880 [b]
1981	9,121	1,574	10,695	1,218	205	1,423	12,118	12,845 [b]
1982	8,867	1,539	10,406	1,095	151	1,246	11,652	12,785 [c]
1983	8,963	1,607	10,570	1,003	155	1,158	11,728	12,825 [c]
1984	8,908	1,606	10,514	960	171	1,131	11,645	12,865 [c]
1985	8,793	1,831	10,624	969	162	1,131	11,755	
1986	8,629	1,753	10,382	902	163	1,065	11,447	
1987	9,304	1,877	11,181	1,039	209	1,248	12,429	13,035 [d]
1988	9,534	1,795	11,329	1,092	175	1,267	12,596	13,465 [d]
1989	9,529	1,696	11,225	1,117	170	1,287	12,512	14,190 [d]
1990	9,920	1,847	11,767	1,345	212	1,557	13,324	13,794 [e]
1991	10,481	1,933	12,414	1,473	280	1,753	14,167	14,268 [e]
1992	10,352	1,888	12,240	1,337	313	1,650	13,890	
1993	11,802	2,223	14,025	-	-	-	14,025	14,742 [e]

NOTES: ADAMHA merged with NIH as of October 1, 1992. Data not currently available for ADAMHA for the years 1976 through 1980. Total number of trainees and fellows for NIH and ADAMHA combined for 1994 is an estimate. Positions recommended by NRC from 1976, 1978, 1981 and 1985 reports included recommendations for health services research.

a 1976 NRC Report.
b 1978 NRC Report.
c 1981 NRC Report.
d 1985 NRC Report.
e 1989 NRC Report as modified by NIH from "full-time equivalent positions" to "full-time training positions".

SOURCE: National Institutes of Health, special tabulations, 1993.

ing grants—being allowed to select trainees without prior review by the National Institutes of Health. Thus, in the 1950s, the basic pattern of the "institutional training grant" was developed. Over the years, the expansion of the national biomedical and behavioral research effort and the attendant demand for scientists and teachers led the NIH training effort to expand even further. By 1969, the number of trainees supported under the original training authority reached 16,000 in that year alone:

> By 1971, NIH training grants and fellowships supported or assisted 37.5 percent of the nation's full-time graduate students in the medical sciences and 21 percent in the life sciences. However, in its presentation of the fiscal year (FY) 1974 budget, the administration made an attempt to eliminate the award of all new training and fellowship grants (Lenfant, 1989).

Congress responded by creating a new training authority: the National Research Service Award Act of 1974 (P.L. 93-348). With this act, Congress established a new program of support for advanced study in the biomedical and behavioral sciences. This program differs from other programs of federal support in important ways. Perhaps most important is the notion that the National Research Service Award (NRSA) augments federal support for "graduate research assistance" by restricting awards to "only those subject areas for which there is a need for personnel". Thus, the NRSA program did not spring full-blown in 1974; it represented a dramatic new direction in a long history of federal support for the training of health scientists.

The NRSA Program

When the National Research Service Awards program was established in 1974, a "novel element" (NRC, 1975) was introduced into federal programs of support: the legislation stipulated that these awards should be restricted to subject areas for which there is a need for personnel. Subsequently, amendments to the National Research Act of 1974 and administrative reorganization led to the addition of such fields of advanced research preparation as nursing research and primary care research.

In 1976, the National Institutes of Health and the Alcohol, Drug Abuse and Mental Health Administration (ADAMHA) together provided 11,550 research training positions through the NRSA program (Table A-1).[2] The program peaked initially at 12,830 awards (in FY 1978) and declined to a low of 11,450 awards in 1986. Since 1986, the program has expanded to its current estimated level of 14,000 awards in FY 1993. Recommendations from the National Research Council throughout this period have generally called for a larger program of support than that provided by the U.S. Congress, although certain detailed recommendations have often been adopted by NIH/ADAMHA.[3]

Of the 13,000 NRSA positions that were awarded in FY 1992 by NIH and ADAMHA, the great preponderance were as trainees on NRSA institutional training grants (84 percent) (Figure A-1). Only 2,200 or 16 percent of the awards were made as fellowships.[4]

Predoctoral fellowship support was restricted to just over 500 recipients and these primarily in the behavioral sciences and health services research. At the postdoctoral level fellowships represent about one-quarter of all awards (Figure A-2).

The overall budget for NRSA support has grown in real or current dollars (Figure A-3) but remained steady in constant dollars. As a share of the NIH/ADAMHA R&D budget, NRSA support has declined from a high of about 9 percent in 1980 to about 5 percent in 1993 (Figure A-4).

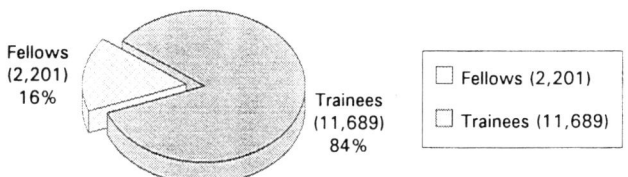

FIGURE A-1 Number of research training positions on NIH and ADAMHA fellowships and training grants, FY 1992. SOURCE: National Institutes of Health, special tabulations, 1993.

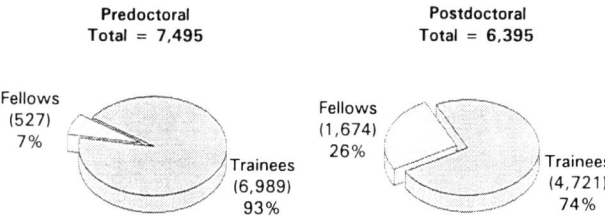

FIGURE A-2 Number of fellows and trainees as a proportion of all predoctoral and postdoctoral NIH and ADAMHA training positions, FY 1992. SOURCE: National Institutes of Health, special tabulations, 1993.

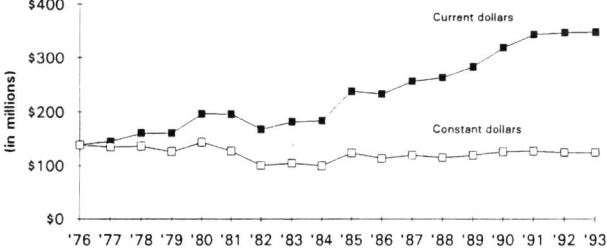

FIGURE A-3 NIH and ADAMHA research training budget, FY 1976-1993. NOTE: Current dollars are estimated. Constant dollars are based on the Biomedical Research and Development Price Index. SOURCE: National Institutes of Health, special tabulations, 1993.

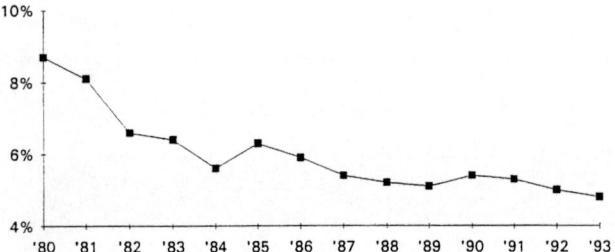

FIGURE A-4 NIH and ADAMHA training budget as a percentage of the budget for extramural research and development grants (estimates). SOURCE: National Institutes of Health, special tabulations, 1993.

ESTABLISHING THE SCOPE OF THE NRSA PROGRAM

The National Research Council has played an active role in advising the National Institutes of Health and the U.S. Congress on the appropriate mix of NRSA support given national needs for research scientists and employment opportunities. The NRC has provided this advice since the enactment of the National Research Act of 1974. Nine reports were issued between 1975 and 1989.

The 1975 report provided definitions for the key concepts basic to this study—training grants, fellowships, institutional support, and predoctoral and postdoctoral training—and discussed their relationship with the quality of biomedical and behavioral research conducted in this country. A short history of the relevant federally supported programs was provided along with a summary of career outcomes of former trainees and fellows who participated in them.

The committee's second report (1976) assessed the current academic labor market and near-term outlook for biomedical and behavioral scientists. In most of these fields, the committee found that an ample supply of Ph.D.s was available. In fact, because the rate of growth in biomedical and behavioral research and development (R&D) expenditures had slowed perceptibly since 1968, and because college enrollments were expected to stabilize by 1980 while Ph.D. production continued at a high level, the committee concluded that a slower rate of growth in labor force in these fields was advisable. Accordingly, the committee recommended a modest reduction in the number of federally supported predoctoral students in the basic biomedical and behavioral areas.

Postdoctoral support, the committee believed, should be held constant in the basic biomedical sciences and increased in other areas. In the behavioral sciences, the recommended shift to predominantly postdoctoral training represented a significant reorientation of federal support and graduate training patterns in this area. This recommendation was developed partly in response to the growing need for more specialized investigators capable of dealing with the increasingly complex research questions in the area of behavior and health. On the other hand, the clinical sciences area was seen as needing increased support to help stimulate the flow of M.D.s into clinical research careers. These initial recommendations were intended to remain in effect until the committee's impressions about the market could be confirmed or modified by further analyses and additional data.

In 1977, the committee found evidence that newly trained biomedical and behavioral Ph.D.s were encountering increasing difficulty in obtaining permanent faculty positions. The number of these Ph.D.s on postdoctoral appointments (which the committee considers to be temporary positions) had been rising at a rate of over 13 percent per year between 1972 and 1975 in the biomedical sciences. Furthermore, the committee's 1977 Survey of Recent Doctorate Recipients showed that more than 40 percent of these postdoctoral appointees in biomedical fields had prolonged their appointments because they could not find suitable employment. These indications of a tight job market facing new Ph.D.s in these fields prompted the committee to recommend an additional 10 percent reduction from the number of predoctoral trainees in the biomedical sciences supported by the federal government in 1976. The postdoctoral recommendation was unchanged.

Certain fields within the basic biomedical sciences exhibited evidence of better-than-average employment prospects and were cited as exceptions to the recommendation for reduced predoctoral support. The fields of biostatistics/biomathematics and epidemiology showed no postdoctoral holding pattern and appeared to be attracting people from closely related fields, such as statistics, that are outside the biomedical sciences. For these fields, the committee recommended no reduction in predoctoral support levels.

In its 1977 report, the committee presented for the first time a systematic treatment of health services research training needs, providing a definition for this emerging research area and a preliminary list of training difficulties that face it. In addition to calling for a continued expansion of mental health services research training, primarily at the predoctoral level through the programs of Alcohol, Drug Abuse, and Mental Health Administration (ADAMHA), the committee called for an extension of the NRSA authority to permit training in the general area of health services research especially through the programs once provided by the National Center for Health Services Research (NCHSR).

Nursing research training was officially brought under the purview of the study by amendments made to the NRSA Act in 1976 (the Health Research and Health Services Amendments of 1976 or P.L. 94-278). In its 1977 report, the committee provided the results of its survey of nurses who had completed their doctoral training between 1971 and 1975. The findings suggested that opportunities for employment for doctorally trained nurses was favorable, and led the committee to suggest an expansion of research training support, predominantly at the predoctoral level.

The committee's 1977 report also discussed the issues of mid-career training and the participation of women and minorities in biomedical and behavioral research; the administrative problems of the three-year limit on awards, the payback provision, announcement fields, and multidisciplinary awards; the education and training process by which most biomedical and behavioral scientists are produced; and the importance of federal support in sustaining the research training system.

Most training grant programs were originally focused on the apparent need for increasing the number of well-trained research personnel. However, in developing a stable continuing policy for government support of training programs in the biomedical sciences, the 1978 NRC committee concluded that it is essential to consider other effects that may be less obvious than the contribution of mere numbers. Many experienced observers believe, for example, that training grants have been just as important in improving the quality of training as in providing for increased numbers.

The 1978 study committee identified four important uses of the institutional training grant: first, one of the most important uses of training grant funds is to provide research equipment and supplies for use by the trainees. Research training is unlike many other forms of education in that it cannot be learned solely from books. Much biomedical research depends also upon the availability of specialized apparatus, costing in the tens to hundreds of thousands of dollars.

> Many of these instruments require special training for their use, and it is the custom in good training laboratories to assign a high level technician to protect the apparatus from misuse and train the graduate students and visiting investigators in its proper handling. Such personnel are often at least partially paid from training grants and certainly play an essential role in the training process (NRC, 1978).

Second, training grants have almost certainly improved the quality of training by providing a portion of the salaries for additional faculty members. One of the major purposes of training grants has been to encourage interdepartmental training programs.

> The field of genetics provides an excellent example. In many institutions the geneticists may be found in several departments—plant geneticists in the botany department, animal geneticists in the zoology department, insect geneticists in the department of entomology, bacterial geneticists in the department of microbiology, and medical geneticists in the medical school—and in universities with an agricultural college, they may be found additionally in the departments of agronomy and plant breeding. In many institutions, training grants have served to bring such scattered teachers together to provide broad training to graduate students and postdoctoral fellows in important fields that transcend departmental boundaries. More often than not, however, some important disciplines may be missing, and training grant funds may be used to fill the gap on either a permanent or visiting basis (NRC, 1978).

The need for such additions to faculty is particularly important in rapidly advancing fields.

Third, training grants contribute to excellence simply by providing an increased number of graduate students to a high-quality department. By careful adjustments of such support, a more equitable distribution of students may be effected without any net overall increase in numbers.

Fourth, there has been so much discussion, both among the public and in the Congress itself, about improving scientific communication, that perhaps one need only mention the importance of training grants in providing for the purchase of essential printed materials and forwarding the information communication which is a critical part of the scientific process.

Some special issues were addressed by the committee during the 1980s. For example, it explored the value of the training system in the biomedical sciences and also the attraction of women and minorities to the biomedical sciences.

Value of the Training System

In the 1983 report, the committee explored the value of NIH traineeships and fellowships by surveying the careers of past recipients. As expected, those supported by these highly coveted competitive awards had achieved an admirable level of success. In general, they attained their degrees in less time, won postdoctoral awards more often, showed greater research productivity and experienced less difficulty in the job market than other scientists. Clearly, the committee felt, the awards programs contributed substantially to developing a cadre of highly capable investigators.

Attraction of Minorities to Careers in Biomedical Research

The Minority Access to Research Careers (MARC) program was created by the National Institute of General Medical Sciences (NIGMS) in 1977 to attract talented minority students to the biomedical sciences. In the 1985 report, at the suggestion of NIGMS, the committee evaluated the MARC Honors Undergraduate Research Training Program. (See also Garrison, et al., 1985.) The committee reported that the program provides special training and research opportunities to selected juniors and seniors at colleges and universities with substantial minority enrollments. The 800 alumni of the program had achieved an excellent record of success, including a number of publications by undergraduate trainees. More than 76 percent of MARC trainees went on to some form of graduate training, and nearly half pursued doctorates. Although most pursued medical or dental degrees, more than a third of the trainees responding foresaw careers in research. All but a handful expected to pursue careers related to science or engineering.

Attraction of Women to Careers in Biomedical Research

The 1989 report expanded the committee's consideration of groups underrepresented in the research pool to include women. Although the number of women receiving Ph.D.s had increased more rapidly than any racial or ethnic minority except Asians, women remained markedly underrepresented among full-time research scientists. Non-Asian racial minorities showed even greater underrepresentation, by factors as great as 6 or 7. Such underutilization of the nation's pool of talent seemed particularly regrettable in light of the dramatic improvement in the market for biomedical researchers, driven by rising demand in industry. The committee explored various possible reasons for these discrepancies.

EVALUATING THE NRSA PROGRAM

The National Research Service Award Act of 1974 poses questions of program outcome as part of the continuing study of national needs. Here the focus is on knowing what happens to awardees (e.g., Are they engaged in health research careers?). Questions concerning program effectiveness also are implied in the legislative authority. During 20 years of study, NRC committees assessing national needs in this area have looked occasionally at the matter of career outcomes and questions of program effectiveness. Chief findings are summarized in the pages that follow.

Predoctoral Training for Ph.D.s

Three major studies examined outcomes associated with NRSA-sponsored predoctoral training (Coggeshall and Brown, 1984; National Research Council, 1976, 1977). In general, the results indicated that NIH awardees distinctly outperformed their comparison groups in terms of greater involvement in research (e.g., receipt of additional postdoctoral research training, time spent in research, and grant application/award activity). These individuals also had somewhat better track records in carrying out high quality research (as measured by citations). Similar to the results of previous studies on the determinants of academic careers (Long et al., 1979; McGinnis and Long, 1988), awardees did not experience any greater success in locating academic employment, once prestige of doctoral institution had been controlled.

A more recent study (Coggeshall and Brown, 1984) of NIH predoctoral awards also attempted to, at least partially, control for the heterogeneity of training experiences and selectivity. Looking at those individuals who received their Ph.D.s in the biomedical sciences between 1967 and 1981, three study groups were compared: (1) those who received at least 9 months of NIH predoctoral support; (2) those who earned their degree from the same departments as the first group but who received 0-8 months of NIH support; and (3) those who graduated from departments that did not have NIH training funds. This strategy permitted two important considerations: (1) those departments receiving NIH funds, often the top-ranked departments in the biomedical sciences, apply the same criteria to accept students, and thus their source of predoctoral support, may be more similar in terms of individual differences (e.g., abilities); and (2) that students who are in departments with NIH funding programs but who are not supported by these funds for an extended length of time may benefit from certain resources accruing to NIH-supported departments.

Postdoctoral Research Training for Ph.D.s

Three major studies have focused on identifying the outcomes of NRSA-supported postdoctoral training, primarily those of biomedical scientists. In general, those with postdoctoral training, regardless of sponsor, outperformed on all measures compared to those who were supported for their predoctoral education but who did not choose to pursue additional postdoctoral study.

More recent examinations of NIH postdoctoral training in the biomedical sciences have been carried out for 1967-1977 Ph.D. recipients (NIH, 1986) and for 1961, 1966, 1971, and 1976 Ph.D. recipients in the biomedical sciences (Garrison and Brown, 1986). Here the major comparison groups were (1) NIH postdoctoral trainees and fellows, (2) Ph.D.s who had likely received postdoctoral training from other sponsors, and (3) those who reported no plans for postdoctoral study at the time they received their degree. Substantial differences emerged between NIH postdoctoral awardees and those who indicated no plans for postdoctoral study; for example, Garrison and Brown (1986) found that NIH awardees were three times as likely as the "no plans" group to have applied for NIH/ADAMHA research grants (56.9 percent versus 19.6 percent) and four times as likely to have been awarded a grant (40.0 percent versus 9.2 percent). This latter difference was reduced somewhat when only those who applied for grants were considered (70.3 percent of NIH awardees versus 47.1 percent of "no plans" groups). They also were more likely than those with no postdoctoral training to have obtained faculty appointments 8-9 years after the Ph.D. (66.7 percent versus 52.7 percent) and, depending on the specific cohort examined, to have published more articles and received more citations per article. A study by NIH (1986) revealed similar findings in terms of academic employment and research funding activity.

As Garrison and Brown (1986) found, NIH awardees continued to outperform in terms of grant application activity those individuals whose postdoctoral training was supported via another source (56.9 percent versus 34.5 percent). Also, they were more likely to have been awarded a grant (40.0 percent versus 22.3 percent). This disparity decreased substantially, however, when considering only those applying for such grants (70.3 percent versus 64.8 percent). There did appear to be some advantage in terms of academic employment; the percentage obtaining a faculty position was 66.7 percent for NIH awardees as compared to 56.7 percent for those with other types of postdoctoral training, but consonant with previous research (McGinnis et al., 1982), this relationship could be primarily accounted for by other factors (e.g., prestige of doctoral institution). Similar results were reported by NIH (1986).

Postdoctoral Training for M.D.s

The role of postdoctoral training for M.D.s was examined by the three studies discussed in the preceding paragraphs. However, the difficulty in interpreting the results—resulting from problems encountered in drawing comparison groups resembling in both orientations and experiences M.D.s with NRSA-supported, postdoctoral research training—is exacerbated by the fact that the vast majority of physicians do not follow research careers. In addition, identifying reasonable comparison groups in these retrospective studies is further complicated by the fact that existing databases for physicians typically are less complete than those for Ph.D. recipients.

Differences between M.D.s with postdoctoral appointments and those without postdoctoral training, some of which appear to be substantial, were found by the National Research Council (1976) for certain outcomes: employment in medical schools and universities (40.9 percent versus 7.4 percent, respectively); the average amount of time reported in conducting research (10.6 percent versus 2.6 percent); and numbers of publications and citations (e.g., 586.6 citations versus 10.3 citations per person for M.D.s aged 41-50). By the use of additional comparison groups, a strong relationship between the existence and length of formal research training and outcomes also appeared—a relationship that has been supported by analyses of more recent trainees (Levey et al., 1988; Sherman, 1983a, 1983b, 1989). In addition to the M.D. groups specified above, two other groups were identified: individuals who had earned both an M.D. and a Ph.D. and who had or had not received postdoctoral training. With the exception of average time spent in research, the results showed a ranking among these groups in line with the amount of research training received. For example, the proportions employed in academic settings were 67.5 percent for M.D./Ph.D.s with postdoctoral appointments, 60.4 percent for M.D./Ph.D.s who did not pursue postdoctoral study, 40.9 percent for M.D.s who had NIH-supported postdoctoral appointments, and 7.4 percent for M.D.s with neither a Ph.D. nor postdoctoral training. On each of the four measures used in the study, the performance of M.D./Ph.D.s, regardless of whether they had been engaged in postdoctoral study, was higher than for those M.D.s who did not possess a Ph.D.

The two remaining studies tried to draw comparison groups that addressed in some way selectivity issues. Rather than looking only at *all* M.D.s without postdoctoral training, Garrison and Brown (1986) also identified another group of M.D.s who received their degree in 1965 or 1974, who reported their primary activities to be "research" or "training," but who had not received postdoctoral research training. Looking at 1974 M.D.s only, there were differences between this group and NIH postdoctoral trainees and fellows. For example, those M.D.s with NIH-supported postdoctoral training also were slightly more likely to have applied for NIH/ADAMHA research grants (18.6 percent versus 12.0 percent) and subsequently been awarded funding (8.7 percent versus 5.5 percent).

A comparison of these outcomes between M.D.s who had NIH postdoctoral fellowships and those who had unsuccessfully *applied* for these fellowships was performed by the NIH (1986). Although both this study and the Garrison and Brown (1986) study demonstrated that NIH fellows comprise a small and select group of M.D.s with NIH postdoctoral awards, this comparison is instructive, although still equivocal, in that it attempts to address some issues of selectivity. Looking at 1986 and 1971 M.D. recipients, the National Institutes of Health found that NIH *fellows* consistently outperformed their unsuccessful applicant counterparts in terms of medical school faculty appointments (65.1 percent versus 43.5 percent) and NIH/ADAMHA application activity (27.4 percent versus 19.4 percent). Of those who applied for grants, 59.1 percent of the fellows versus 33.3 percent of the unsuccessful fellow applicants received an award.

In general, all of the previously described studies on predoctoral and postdoctoral training have contributed to our knowledge about certain accomplishments of NRSA awardees. Because of unresolved problems with selectivity and heterogeneity of training experiences, however, no definitive evaluation study has yet been undertaken to provide strong evidence for the absolute effect of NRSA training on career outcome.

APPENDIX A

NOTES

1. Material for this section has been drawn from a variety of sources, including work commissioned by the study committee in 1992 and provided by Ms. Judith Grumstrup-Scott and Ms. Beryl Benderly. Sections have also been drawn from previous NRC study committee reports, most notably that of the 1978 study committee. The last section on evaluation issues is drawn from a paper by Dr. Georgine Pion commissioned by the 1989 NRC study committee.

2. In October 1992, the research components of the three institutes of ADAMHA joined the National Institutes of Health.

3. For example, earlier study committees called for a shift from predoctoral to postdoctoral support in the behavioral sciences, which has occurred.

4. The relative diminution of the fellowship mechanism within the NRSA program is an interesting feature of the history of the NRSA program. The 1974 authorization specified that not less than 25 percent of total support should be directed to fellowship support; the fraction has declined to 15 percent in recent years.

REFERENCES

Ahrens, E.H., Jr.
    1992   *The Crisis in Clinical Research: Overcoming Institutional Obstacles.* New York: Oxford University Press.

Coggeshall, P.E. and P.W. Brown
    1984   *The Career Achievements of NIH Predoctoral Trainees and Fellows.* Washington, D.C.: National Academy Press.

Garrison, H.H. and P.W. Brown
    1986   *Career Achievements of NIH Postdoctoral Trainees and Fellows.* Washington, D.C.: National Academy Press.

Garrison, H.H., P.W. Brown and R.W. Hill
    1985   *Minority Access to Research Careers: An Evaluation of the Honors Undergraduate Research Training Program.* Washington, D.C.: National Academy Press.

Lenfant, C.
    1989   *Review of the National Institutes of Health Biomedical Research Training Programs.* Bethesda, Maryland: NIH.

Levey, G.S., C.R. Sherman, N.O. Gentile, L.J. Hough, T.H. Dial, and P. Jolly
    1988   Postdoctoral research training of full-time faculty in academic departments of medicine. *Annals of Internal Medicine* 109:414-418.

Long, J.S., P.D. Allison, and R. McGinnis
    1979   Entrance into the academic career. *American Sociological Review* 44:816-830.

McGinnis, R., P.D. Allison, and J.S. Long
    1982   Postdoctoral training in bioscience: Allocation and outcomes. *Social Forces* 60:701-723.

McGinnis, R. and J.S. Long
    1988   Entry into academia: Effects of stratification, geography, and ecology. In *Academic Labor Markets and Careers*, ed. D.W. Breneman and T.I.K. Youn. New York: Falmer Press.

National Institutes of Health (NIH)
    1986   *Effects of the National Research Service Award Program on Biomedical Research and Teaching Careers.* Bethesda, MD: NIH.

National Research Council (NRC)
    1975   *Personnel Needs and Training for Biomedical and Behavioral Research.* Washington, D.C.: National Academy of Sciences.
    1976   *Research Training and Career Patterns of Bioscientists: The Training Programs of the National Institutes of Health.* Washington, D.C.: National Academy of Sciences.
    1977   *Personnel Needs and Training for Biomedical and Behavioral Research.* Washington, D.C.: National Academy of Sciences.
    1978   *Personnel Needs and Training for Biomedical and Behavioral Research.* Washington, D.C.: National Academy of Sciences.

Sherman, C.R.
    1983a   *Notes on the NIH Role in Support of Postdoctoral Research Training of Two Groups of Physicians.* (Available from Charles Sherman, NIH, 9000 Rockville Pike, Bethesda, MD 20092.)
    1983b   Training and Manpower Development. Presentation at the Meeting of the Advisory Committee to the Director, NIH, Bethesda, MD.
    1989   *The NIH Role in the Training of Individual Physician Faculty: A Supplementary Analysis.* (Available from Charles Sherman, NIH, 9000 Rockville Pike, Bethesda, MD 20092.)

APPENDIX B

CLASSIFICATION OF FIELDS

BASIC BIOMEDICAL SCIENCES

Anatomy
Bacteriology
Biochemistry
Biomathematics
Biomedical Engineering
Biometrics and Biostatistics
Biophysics
Cell Biology/Cytology
Dentistry
Embryology
Endocrinology
Environmental Sciences
Environmental Health
Epidemiology
Food Science and Technology
General Biological Sciences
General Health, Medical Sciences Hospital Administration
Human and Animal Genetics
Human and Animal Pathology
Human and Animal Pharmacology
Human and Animal Physiology
Immunology
Medicine and Surgery
Microbiology
Molecular Biology
Neurosciences
Nutritional Sciences/Dietetics
Optometry, Ophthalmology
Other Biological Sciences
Other Health/Medical Sciences
Parasitology
Pharmaceutical Chemistry
Pharmacy
Public Health
Toxicology
Veterinary Medicine
Zoology

BEHAVIORAL SCIENCES

Clinical Psychology

Clinical Psychology
Counseling and Guidance
School

Non-Clinical Psychology

Behavior/Ethology
Cognitive
Comparative
Developmental and Child
Educational
Experimental
Human Engineering
Industrial and Organizational
Other Psychology
Personality
Physiological
Psychometrics
Quantitative
Social

Other Behavioral Sciences

Anthropology
Audiology and Speech Pathology
Sociology

APPENDIX B

CLINICAL SCIENCES

Allergy
Anesthesiology
Clinical Psychology
Clinical Dentistry
Geriatrics
Internal Medicine
Neurology
Neuropsychiatry
Obstetrics and Gynecology
Ophthalmology
Other Clinical Medicine
Otorhinolaryngology
Pediatrics
Preventive Medicine
Psychiatry
Radiology
Social Work
Surgery
Veterinary Medicine

ORAL HEALTH RESEARCH

(See Chapter 6)

NURSING RESEARCH

(See Chapter 7)

HEALTH SERVICES RESEARCH

(See Chapter 8)

APPENDIX C

PUBLIC HEARING ON NATIONAL NEEDS FOR BIOMEDICAL AND BEHAVIORAL RESEARCH PERSONNEL

May 3, 1993
Washington, D.C.

PROGRAM

A continuing goal of our national research effort is to sustain the quality of biomedical and behavioral research. To achieve this goal we must maintain research training environments of high quality and sufficient stability to assure the future availability of skilled research personnel.

The National Research Service Awards Act of 1974 established a Federal program of predoctoral and postdoctoral training support to meet national needs for biomedical and behavioral scientists. At the same time, the Act requested that the National Academy of Sciences undertake a continuing study of personnel needs in this area and that they report on a regular basis to the U.S. Congress, the National Institutes of Health, and related agencies regarding future training needs in this area.

The National Research Council, the operating arm of the National Academy of Sciences, recently established the Committee on National Needs for Biomedical and Behavioral Research Personnel to conduct a study of this topic. Following its first meeting, the Committee arranged to convene a one-day Public Hearing to explore four overarching questions:

1. What is the most significant challenge we face today in the United States for maintaining an adequate supply of qualified scientists to sustain and advance health research?

2. What improvements might be made in the National Research Service Awards program to assure a continuing supply of skilled investigators in the biomedical and behavioral sciences in the coming years?

3. What steps might be taken to improve the effectiveness of the National Research Service Award program in recruiting women and minorities into scientific careers?

4. What features of the National Research Service Award training grant might be strengthened to assure the maintenance of high quality research training environments?

The Committee is also aware of significant changes that have come about in employment opportunities for bioscientists in industry and other types of nontraditional research settings. The Public Hearing provides a forum for exploring this issue as well.

AGENDA

8:00 a.m.	*Registration/Coffee*
8:45 a.m.	*Welcome and Introduction of Committee/Staff* • Ira Hirsh and John D. Stobo
9:00 a.m.	Susan Gerbi, Brown University
9:10 a.m.	David Brautigan, American Society for Biochemistry and Molecular Biology
9:20 a.m.	Terry Ann Krulwich, The Mount Sinai School of Medicine
9:30 a.m.	*Panel (Fellows/Trainees)* • Julie Fielding, Brigham and Women's Hospital • Vincent LiCata, University of Minnesota • Ora Weisz, The Johns Hopkins University
10:00 a.m.	*General Discussion*
10:15 a.m.	**BREAK**
10:30 a.m.	Thomas Malone, American Association of Medical Colleges
10:40 a.m.	Bryan Marshall, University of Pennsylvania
10:50 a.m.	George Kimmich, University of Rochester

APPENDIX C

11:00 a.m.	*Panel (Fellows/Trainees)* • Miyuki Yamaguchi, Duke University • JoAnne Pohl, The University of Michigan
11:30 a.m.	*General Discussion*
12 Noon	LUNCH
1:30 p.m.	Susan Person, Consortium of Social Science Associations
1:40 p.m.	James Jones for Wayne Camara, American Psychological Association
1:50 p.m.	John McCormick speaking on behalf of Judson Sheridan, University of Missouri
2:00 p.m.	*Panel (Fellows/Trainees)* • Barton Giddings, Whitehead Institute for Biomedical Research • Phillip Cozzi, University of Chicago • Pat McCloskey, University of North Carolina at Chapel Hill
2:30 p.m.	*General Discussion*
2:45 p.m.	BREAK
3:00 p.m.	Gail Cassell, American Society for Microbiology
3:10 p.m.	Herbert B. Silber, San Jose State University
3:20 p.m.	Daniel Linzer, Northwestern University
3:30 p.m.	Ada K. Jacox, American Nurses Association
3:40 p.m.	*Panel (Fellows/Trainees)* • Thomas Meyer, Columbia University • Carey Lumeng, University of Michigan
4:00 p.m.	*Panel (Training Grant Directors)* • Irwin Sandler, Arizona State University • Lee Goldman, Harvard University • Peter Shank, Brown University • Page Morahan, Medical College of Pennsylvania
4:40 p.m.	*Panel (Professional Society Representatives)* • Cornelius Pings, Association of American Universities • Alan Kraut, American Psychological Society • Harold Slavkin, American Association for Dental Research • Dominick Purpura, Society for Neuroscience
5:20 p.m.	Elizabeth Jones, Carnegie Mellon University
5:30 p.m.	J. Fredrick Dice speaking on behalf of Irwin Arias, Tufts University
5:40 p.m.	Homayoun Kazemi, Massachusetts General Hospital
5:50 p.m.	Richard Grand, Tufts University
6:00 p.m.	*General Discussion*
6:30 p.m.	ADJOURNMENT

RESPONDENTS TO SOLICITATION FOR COMMENTS

Graduate Deans

Catholic University of America
Mary Jean Flaherty
School of Nursing

Columbia University
Roger Bagnall
Graduate School of Arts and Sciences

Cornell University
Douglas McGregor
College of Veterinary Medicine

Harvard University
Christoph Wolff
Graduate School of Arts and Sciences

Rockefeller University
Bruce McEwen
Graduate and Postgraduate Studies

Rutgers University-New Brunswick
Joseph A. Potenza
Graduate School

University of California-Los Angeles
Robin Fisher
Graduate Division

University of Chicago
Jeffrey Slovak
Graduate School

University of Illinois at Urbana-Champaign
Chester Gardner
Graduate College

University of Iowa
Leslie B. Sims
Graduate College

University of Missouri-Columbia
Judson Sheridan
Graduate School

Grant Directors

Robert Ader
Center for Advanced Study in the Behavioral Sciences
Stanford, CA

Irwin M. Arias
Tufts University
Department of Physiology

Marilyn J. Aten
University of Rochester Medical Center
School of Nursing

David M. Austin
University of Texas at Austin
School of Social Work

John L. Azevedo, Jr.
East Carolina University School of Medicine
Department of Biochemistry

Donald Bartlett, Jr.
Dartmouth Medical School
Department of Physiology

Howard Baum
Massachusetts General Hospital
Neuroendocrine Clinical Center

Robert Blank
The Rockefeller University

Daniel F. Bogenhagen
State University of New York at Stony Brook,
Department of Pharmacology

Selwyn A. Broitman
Boston University School of Medicine
Department of Microbiology

Nicholas Cohen
University of Rochester School of Medicine and
Dentistry, Department of Microbiology and Immunology

Shelley A. Cole
Southwest Foundation for Biomedical Research
Department of Genetics

Robert E. Cone
University of Connecticut Health Center
Department of Pathology

Jose R. Criado
The Scripps Research Institute
Department of Neuropharmacology

Bruce P. Dohrenwend
College of Physicians & Surgeons at Columbia University,
Social Psychiatry Research Unit

Emanuel Donchin
University of Illinois at Urbana-Champaign
Department of Psychology

Brian R. Duling & Robert M. Berne
University of Virginia Health Sciences Center,
Department of Physiology

Floyd Dunn
University of Illinois at Urbana-Champaign Bioacoustics
Research Laboratory

Lise S. Eliot
Baylor College of Medicine
Division of Neuroscience

Barbara L. Finlay
Cornell University
Department of Psychology

Frances M. Finn
University of Pittsburgh School of Medicine,
Protein Research Laboratory

Elizabeth A. Franz
University of California, Berkeley
Department of Psychology

Robert N. Golden
The University of North Carolina at Chapel Hill
Department of Psychiatry

Lowell A. Goldsmith
University of Rochester Medical Center
Department of Dermatology

APPENDIX C

David Hamerman
Albert Einstein College of Medicine
Resnick Gerontology Center

Robert I. Handin
Harvard Medical School/Brigham & Women's Hospital,
Hematology-Oncology Division

George A. Hedge
West Virginia University Health Sciences Center,
School of Medicine Research and Graduate Studies

Arthur Horwich
Yale University School of Medicine
Department of Genetics

William P. Jencks
Brandeis University
Graduate Department of Biochemistry

Richard T. Johnson
Johns Hopkins University
Department of Neurology

Agnes B. Kane
Brown University
Division of Biology and Medicine
Department of Pathology and Laboratory Medicine

Ira R. Katz
Medical College of Pennsylvania
Eastern Pennsylvania Psychiatric Institute

Ann R. Kennedy
University of Pennsylvania
School of Medicine
Department of Radiation Oncology

John M. Kirkwood
Pittsburgh Cancer Institute
Melanoma Center

David M. Levine
The Johns Hopkins University School of Medicine
Division of Internal Medicine

Stephen J. Lippard
Massachusetts Institute of Technology
Department of Chemistry

Ian G. Macara
The University of Vermont
Department of Pathology

Walter Makous
University of Rochester College of Arts and Science
Center for Visual Science

Kenneth G. Mann
The University of Vermont
College of Medicine
Department of Biochemistry

Bryan E. Marshall
University of Pennsylvania
Department of Anesthesia

Richard H. Masland
Harvard Medical School
Massachusetts General Hospital

Richard E. McCarty
Johns Hopkins University
Department of Biology

D. Kent Morest
The University of Connecticut Health Center
Center for Neurological Sciences

Howard E. Mossberg
The University of Kansas
Office of the Vice Chancellor for Research,
Graduate Studies, & Public Service

Nancy Elsa Mueller
Harvard School of Public Health
Department of Epidemiology

J. Michael Mullins
The Catholic University of America
Department of Biology

Eric G. Neilson
University of Pennsylvania Medical Center
Renal-Electrolyte Division

James O'Rourke
The University of Connecticut Health Center,
Vision Immunology Center

Michael C. Phillips
Medical College of Pennsylvania
Department of Biochemistry

Cedric S. Raine
Albert Einstein College of Medicine of Yeshiva University
Department of Pathology/Neuropathology

Katheleen M. Rasmussen
Cornell University
Division of Nutritional Sciences

Charles F. Reynolds
University of Pittsburgh Medical Health Care Division
Clinical Research Training in Psychiatry

Harold R. Roberts
University of North Carolina at Chapel Hill
School of Medicine
Division of Hematology

William N. Rom
New York University Medical Center
Division of Pulmonary and Critical Care Medicine

Neil Ruderman
Boston University Medical Center
Diabetes and Metabolism Unit

Marijane Russell
National Jewish Center for Immunology and
Respiratory Medicine

Robert Snyder
Rutgers University
Joint Graduate Program in Toxicology

Jerry L. Spivak
The Johns Hopkins University School of Medicine
Division of Hematology

Thomas P. Stossel
Harvard Medical School, Brigham and Women's Hospital
Experimental Medicine Division

Jerome F. Strauss, III & Luigi Mastroianni
University of Pennsylvania Medical Center
Department of Obstetrics and Gynecology

Brian L. Strom
University of Pennsyvania Medical Center
Center for Clinical Epidemiology and Biostatistics

Ming T. Tsuang
Harvard School of Public Health
Harvard Program in Psychiatric Epidemiology

Watt W. Webb
Cornell University College of Engineering
School of Applied and Engineering Physics

Gordon H. Williams
Brigham & Women's Hospital, Harvard Medical School
Endocrinology-Hypertension Division

Sankey V. Williams
University of Pennsylvania
Division of General Internal Medicine

William M. Willingham
University of Arkansas at Pine Bluff
Center for Multi-Purpose Research and
Sponsored Programs

Raymond L. Woosley
Georgetown University Medical Center
Department of Pharmacology

Frank C. P. Yin
The Johns Hopkins Hospital
Clayton Heart Center

Michael J. Zigmond
University of Pittsburgh
Department of Behavioral Neuroscience

Professional Associations

American Academy of Nursing
Janet Heinrich

American Association for Dental Research
John J. Clarkson

American Association of Colleges of Nursing
Geraldine Polly Bednash

American Association of Dental Schools
Preston A. Littleton

American Chemical Society
Helen M. Free

American College of Dentists
Sherry Keramidas

American Dental Association
John S. Zapp

American Institute of Biological Sciences
Donald R. Beem

American Nurses Association
Barbara K. Redman

American Psychiatric Association
Harold Alan Pincus

American Psychological Association
Wayne J. Camara

American Psychological Society
Alan G. Kraut

American Society for Biochemistry and
Molecular Biology
Charles C. Hancock

American Speech-Language-Hearing Association
Frederick T. Spahr

Association for Health Services Research
John M. Eisenberg, Roger J. Bulger,
Alice S. Hersh

Association of American Medical Colleges
Robert G. Petersdorf

Association of American Universities
Cornelius J. Pings

Association of American Veterinary Medical Colleges
Ronald A. Wright

International Association for Dental Research, Behavioral
Sciences and Health Services Research Group
Jane A. Weintraub

Oncology Nursing Society
Mel Haberman

Students

Merritt B. Andrus
Harvard University
Department of Chemistry

John M. Armstrong
University of Texas Southwestern Medical Center

Melissa Berhow
Yale University School of Medicine

Daniel Bossut
University of North Carolina at Chapel Hill
Department of Physiology

Susan C. Boynton
Brandeis University
Department of Biology

David L. Brody
Johns Hopkins University Medical School Medical
Scientist Training Program

Rebecca Burwell
The Salk Institute

Miriam E. Cameron
University of Minnesota
School of Nursing

Dominick DePhilippis
University of California, San Francisco
Substance Abuse Research

Anne Harwood
National Jewish Center
Department of Pediatrics

Mark Hearn
University of Washington
Department of Pathology

John A. Koch
Harvard Medical School
Dana-Farber Cancer Institute

Lori A. Loan

Robert J. McDonald
Boston University Medical Center

James E. McDuffie

David E. Morledge
Temple University School of Medicine

Tom Pauly
Purdue University
Biology Department

R.A. Perez-Tamayo
Duke University Medical Center
Department of General and Thoracic Surgery

Kory J. Schuh
The Johns Hopkins University School of Medicine
Behavioral Pharmacology Research Unit

APPENDIX C

Brian Slezak
State University of New York at Buffalo

Jaime Steinsapir
Medical College of Georgia
Department of Physiology and Endocrinology

Samuel W. Straight
University of Rochester
Department of Microbiology and Immunology

Margaret F. Tremwel
University of Florida
Department of Neuroscience

Michael J. Woller
The University of South Carolina School of Medicine
Department of Cell Biology and Neuroscience

Marcia Sydney Zax

APPENDIX D

REPORT CONTRIBUTORS

The following individuals provided information for committee use in their role as consultants and/or authors of commissioned papers:

BERYL BENDERLY, Science writer, Washington, D.C.

FARRELL BLOCH, Independent consultant, Washington, D.C.

SHARON BUSH, Office of Scientific and Engineering Personnel, National Research Council, Washington, D.C.

ROSEMARY CHALK, Commission on Behavioral and Social Sciences and Education, National Research Council, Washington, D.C.

CAROLA EISENBERG, Department of Psychiatry, Harvard University Medical School, Cambridge, Massachusetts (Liaison: NRC/OSEP Committee on Women in Science and Engineering)

DONALD S. FREDERICKSON, Independent consultant, Bethesda, Maryland

JUDITH GRUMSTROP-SCOTT, Science writer, Washington, D.C.

JANET HEINRICH, American Academy of Nursing, Washington, D.C.

ERNEST JAWORSKI, Monsanto Company (Retired) (Liaison: NRC/OSEP Advisory Committee)

LAURA LATHROP, Office of Scientific and Engineering Personnel, National Research Council, Washington, D.C.

MATTHEW MURRAY, Lawrence Berkeley Laboratory, Human Genome Center, Berkeley, California

GEORGINE M. PION, Vanderbilt Institute for Public Policy Studies, Vanderbilt University, Nashville, Tennessee

PETER TIEMEYER, Consultant, RAND, Santa Monica, California

JAMES A. VOYTUK, Office of Scientific and Engineering Personnel, National Research Council, Washington, D.C.

JOSEPH L. ZELIBOR, Commission on Life Sciences, National Research Council, Washington, D.C.

APPENDIX E

SOURCES OF INFORMATION FOR THE NATIONAL RESEARCH SERVICE AWARD

The National Institutes of Health provide support through the National Research Service Awards at the predoctoral and postdoctoral level. At each level the programs are distinguished by whether they are made directly to individuals who use the support at an institution of their choice, or to institutions, who in turn make awards to individuals in their programs. The following is a list of programs by level and type for the National Research Service Awards:

Predoctoral Level

Individual Awards

PREDOCTORAL NATIONAL RESEARCH SERVICE AWARD (F31)

Individual Predoctoral Fellowships for Nursing Research—For research training leading to a doctoral degree in biomedical and behavioral fields relevant to nursing. Information available from the National Institute of Nursing Research.

NRSA Predoctoral Fellowships for Minority Students Awards—Provides support for research training leading to the Ph.D. degree or the M.D./Ph.D. in biological sciences for selected students from underrepresented minority groups. Information available from the National Institute of General Medical Sciences.

MARC Predoctoral Fellowships—Provides support for research training leading to the Ph.D. degree in biological sciences for selected students who are graduates of the MARC Honors Undergraduate Research Training Program. Information available from the National Institute of General Medical Sciences.

MARC National Research Service Award Faculty Fellowship (F34)—Provides opportunity for advanced predoctoral research training of selected faculty at eligible institutions in which student enrollments are drawn substantially from minority groups. Information available from the National Institute of General Medical Sciences.

Institutional Awards

NRSA Institutional Training Grants (T32)—Awarded to eligible institutions to develop or enhance research training opportunities for individuals, selected by the institution, who are training for careers in specified areas of biomedical and behavioral research. Information available from the Office of Grant Inquiries, Division of Research Grants, National Institutes of Health.

Medical Scientist Training Program (MSTP)—Awarded to selective institutions to provide a maximum of 6 years of support for students in an integrated scientific and medical program leading to the combined M.D./Ph.D. Information available from the National Institute of General Medical Sciences.

MARC Honors Undergraduate Research Training Program (T34)—Awarded to selected institutions to support the undergraduate education of minority students who can compete successfully for entry into graduate programs leading to a Ph.D. degree. Information available from the National Institute of General Medical Sciences.

NHLBI Short Term Training for Minority Students Program (T35)—Awarded to selected institutions to provide minority undergraduate students, graduate students, and students in health professional schools short-term exposure to research in areas relevant to cardiovascular, pulmonary, and hematologic diseases. Information available from the Office of Grant Inquiries, Division of Research Grants, National Institutes of Health.

Postdoctoral Level

Individual Awards

National Research Service Award Individual Postdoctoral Fellowship (F32)—Awards are for individuals who have received a Ph.D., M.D., D.D.S., D.V.M., or equivalent degree and wish to receive full time training in areas that reflect national need for biomedical and behavioral research. Information available from the Office of Grant Inquires, Division of Research Grants, National Institutes of Health.

MARC Faculty Fellowships (F34)—Provides opportunity for advanced research training of selected faculty members at eligible institutions in which student enrollments are drawn substantially from minority groups. Information available from the National Institute of General Medical Sciences.

National Research Service Award Senior Postdoctoral Fellowship (F33)—Awards are designed to permit scientists with at least 7 years of postdoctoral research or professional training to make a major change in the direction of their research career, or to broaden their scientific background. Information available from the Office of Grant Inquiries, Division of Research Grants, National Institutes of Health.

Intramural National Research Service Awards Research Training Program (F35)—This program provides opportunities for post-M.D. training at one of the NIH Institutes. Applicants must hold the M.D., D.D.S., D.V.M., or equivalent degree. Information available from the Deputy Director of Intramural Research, National Institutes of Health.

Institutional Awards

National Research Service Award Institutional Research Training Grants (T32)—Awarded to domestic nonprofit private or public educational institutions to develop or enhance research training opportunities for individuals interested in research training beyond the doctorate to prepare for careers in biomedical and behavioral research. Information available from the Office of Grant Inquiries, Division of Research Grants, National Institutes of Health.

NHLBI Minority Institutional Research Training Program (T35)—Awarded to selected institutions to provide minority undergraduate students, graduate students, and students in health professional schools short-term exposure. This program is designed to offer research training at minority institutions in areas relevant to cardiovascular, pulmonary, and hematologic diseases. Information available from the Research Training and Development Branch; Division of Heart and Vascular Diseases; National Heart, Lung, and Blood Institute, National Institutes of Health.

APPENDIX F
DATA TABLES

APPENDIX TABLE F-1 National Support for Health R&D, by Performer and Source of Funds, 1980-92

	1980	1981	1982	1983	1984	1985	1986	1987	1988	1989	1990	1991 (est.)	1992 (est.)
						Millions of current dollars							
Total	7,967	8,739	9,595	10,778	12,159	13,565	14,899	16,940	19,011	20,977	23,076	25,561	28,125
Source of Funds													
Government	5,203	5,413	5,612	6,117	6,887	7,675	7,929	9,037	9,725	10,634	11,422	12,413	13,424
Federal	4,723	4,848	4,970	5,399	6,087	6,791	6,895	7,847	8,425	9,163	9,791	10,711	11,596
NIH	3,182	3,333	3,433	3,789	4,257	4,828	5,005	5,852	6,292	6,778	7,136	7,711	8,423
State and Local	480	564	642	718	800	884	1,034	1,191	1,300	1,471	1,632	1,702	1,823
Industry	2,459	2,998	3,593	4,205	4,765	5,352	6,188	7,103	8,432	9,404	10,634	12,020	13,505
Private nonprofit	305	328	390	456	507	538	782	800	854	939	1,020	1,128	1,196
Howard Hughes[a]	18	20	25	54	79	51	247	183	179	197	215	250	281
Performer													
Government	1,487	1,575	1,669	1,813	1,997	2,140	2,155	2,389	2,590	2,578	2,861	3,300	3,568
Federal	1,284	1,364	1,448	1,577	1,741	1,869	1,848	2,042	2,213	2,161	2,403	2,816	3,049
State and Local	203	211	221	236	256	271	307	347	377	417	458	484	520
Industry[b]	2,249	2,659	3,161	3,668	4,216	4,660	5,293	6,002	6,927	7,901	8,817	9,578	11,006
Higher education[b]	3,005	3,211	3,388	3,779	4,274	4,745	5,320	5,056	6,593	7,238	7,744	8,467	9,173
Private nonprofit[b]	726	751	485	887	976	1,115	1,157	1,352	1,455	1,798	1,886	1,931	2,087
Foreign	499	543	593	631	697	805	975	1,140	1,446	1,462	1,769	2,078	2,291

APPENDIX F

	1980	1981	1982	1983	1984	1985	1986	1987	1988	1989	1990	1991 (est.)	1992 (est.)
Biomedical R&D price index [c]	0.649	0.713	0.774	0.819	0.867	0.911	0.949	1	1.05	1.106	1.166	1.224	1.284
							Millions of 1987 dollars						
Total	12,276	12,257	12,397	13,160	14,024	14,890	15,700	16,940	18,106	18,967	19,791	20,883	21,904
Source of Funds													
Government	8,017	7,592	7,251	7,469	7,943	8,425	8,355	9,037	9,262	9,615	9,796	10,141	10,455
Federal	7,277	6,799	6,421	6,592	7,021	7,454	7,266	7,847	8,024	8,285	8,397	8,751	9,031
NIH	4,903	4,675	4,435	4,626	4,910	5,300	5,274	5,852	5,992	6,128	6,120	6,300	6,560
State and Local	740	791	829	877	923	970	1,090	1,191	1,238	1,330	1,400	1,391	1,420
Industry	3,789	4,205	4,642	5,134	5,496	5,875	6,521	7,103	8,030	8,503	9,120	9,820	10,518
Private nonprofit	470	460	504	557	585	591	824	800	813	849	875	922	931
Howard Hughes [a]	28	28	32	66	91	56	260	183	170	178	184	204	219
Performer													
Government	2,291	2,209	2,156	2,214	2,303	2,349	2,271	2,389	2,467	2,331	2,454	2,696	2,779
Federal	1,978	1,913	1,871	1,926	2,008	2,052	1,947	2,042	2,108	1,954	2,061	2,301	2,375
State and Local	313	296	286	288	295	297	323	347	359	377	393	395	405
Industry [b]	3,465	3,729	4,084	4,479	4,863	5,115	5,577	6,002	6,597	7,144	7,562	7,825	8,572
Higher education [b]	4,630	4,504	4,377	4,614	4,930	5,209	5,606	5,056	6,279	6,544	6,642	6,917	7,144
Private nonprofit [b]	1,119	1,053	627	1,083	1,126	1,224	1,219	1,352	1,386	1,626	1,617	1,578	1,625
Foreign	769	762	766	770	804	884	1,027	1,140	1,377	1,322	1,517	1,698	1,784

a For Howard Hughes Medical Institute, figures are for the direct conduct of biomedical research, and exclude support for scientific career development. Figures for 1985 include only 8 months of operations because of change in fiscal year.
b Includes expenditures for federally funded research and development centers administered by organizations in the respective sectors.
c The NIH Biomedical Research and Development Price Index differs from the implicit price deflator for the Gross National Product.

SOURCE: National Science Board, *Science and Engineering Indicators - 1993*. Washington, D.C.: National Science Foundation, 1994.

APPENDIX TABLE F-2 R&D in the National Institutes of Health: Congressional Action on R&D in the FY 1994 Budget (budget authority in millions of dollars)[a]

	FY 1993 Est.	FY 1994 Requested	FY 1994 Approved	Action by Congress Change from Request Amount	Percent	Change from FY 1993 Amount	Percent
Cancer	1,978.3	2,142.1	2,082.3	-59.8	-2.8	104.0	5.3
Heart, Lung and Blood	1,214.7	1,198.4	1,277.9	79.5	6.6	63.2	5.2
Dental Research	161.1	163.0	169.5	6.5	4.0	8.4	5.2
Diabetes, Digestive, and Kidney Diseases	680.7	677.1	716.5	39.4	5.8	35.8	5.3
Neurological Disorders and Stroke	599.5	590.1	630.7	40.6	6.9	31.2	5.2
Allergy and Infectious Diseases	988.4	1,065.6	1,065.6	0.0	0.0	77.2	7.8
General Medical Sciences	832.2	833.1	875.5	42.4	5.1	43.3	5.2
Child Health & Human Development	527.8	542.4	555.2	12.8	2.4	27.4	5.2
Eye	275.9	272.2	290.3	18.1	6.6	14.4	5.2
Environment Health Sciences	251.2	261.3	264.2	2.9	1.1	13.0	5.2
Aging	399.5	394.2	420.3	26.1	6.6	20.8	5.2
Arthritis & Musculoskeletal & Skin Diseases	212.2	210.4	223.3	12.9	6.1	11.1	5.2
Deafness & Communication Disorder	154.8	153.1	162.8	9.7	6.3	8.0	5.2
Research Resources	312.7	327.9	331.9	4.0	1.2	19.2	6.1
Nursing Research	48.5	49.0	51.0	2.0	4.1	2.5	5.2
Alcoholism and Alcohol Abuse	176.4	173.6	185.6	12.0	6.9	9.2	5.2
Drug Abuse	404.2	407.1	425.2	18.1	4.4	21.0	5.2
Mental Health	583.1	576.0	613.4	37.4	6.5	30.3	5.2
Human Genome	106.1	134.5	128.7	-5.8	-4.3	22.6	21.3
Fogarty International Center	19.7	20.0	21.7	1.7	8.5	2.0	10.2
National Library of Medicine	103.6	133.3	120.0	-13.3	-10.0	16.4	15.8
Office of Director	190.3	234.9	233.6	-1.3	-0.6	43.3	22.8
Buildings and Facilities	108.7	108.7	111.0	2.3	2.1	2.3	2.1
TOTAL, NIH Budget	10,329.6	10,668.0	10,956.2	288.2	2.7	626.6	6.1
Estimated Research Training	-349.0	-355.0	-373.4	-18.4	5.2	-24.4	7.0
Other Non-R&D	-90.6	-105.1	-107.6	-2.5	2.4	-17.0	18.8
TOTAL, NIH R&D	9,890.0	10,207.9	10,475.2	267.3	2.6	585.2	5.9

a Author's estimates. Includes conduct of R&D and R&D facilities.
SOURCE: American Association for the Advancement of Science, *Congressional Action on Research and Development in the FY 1994 Budget*. Washington, D.C.: AAAS, 1993.

APPENDIX TABLE F-3 Age Profile of the U.S. Biomedical Science Work Force by Gender: 1981-1991 (in percent)

Age	1981	1983	1985	1987	1989	1991
Total Biomedical	**64,538**	**66,363**	**75,755**	**81,058**	**88,647**	**91,959**
Under 30	5.35	4.75	3.43	2.36	2.72	2.44
30-34	20.20	18.40	17.96	17.29	15.54	13.06
35-39	24.37	23.36	22.66	20.91	20.20	22.17
40-44	16.51	19.48	21.18	21.94	20.20	19.89
45-49	10.34	11.27	12.80	14.54	17.27	18.40
50-54	9.90	9.02	8.38	8.89	9.56	11.59
55-59	7.05	7.55	7.08	7.06	6.74	6.47
60-64	4.28	4.17	4.29	4.89	5.37	4.01
65-69	1.70	1.61	1.81	1.73	1.91	1.76
70+	0.31	0.39	0.41	0.40	0.50	0.21
Median Age (years)	39.04	39.90	40.42	41.16	41.87	42.11
Total Men	**53,868**	**54,219**	**60,800**	**64,317**	**69,083**	**70,309**
Under 30	4.59	4.24	2.73	1.91	2.31	1.99
30-34	19.01	16.88	16.44	15.33	13.92	11.34
35-39	24.37	22.98	22.03	20.20	19.01	20.74
40-44	16.85	19.95	21.75	22.20	20.06	19.44
45-49	10.83	11.85	13.54	15.45	18.39	19.41
50-54	10.39	9.60	9.00	9.73	10.24	12.87
55-59	7.31	7.89	7.42	7.47	7.37	7.45
60-64	4.58	4.51	4.67	5.33	6.03	4.54
65-69	1.78	1.72	1.97	1.97	2.16	2.01
70+	0.28	0.37	0.47	0.42	0.51	0.22
Median Age (years)	40.10	40.98	41.03	42.33	43.18	43.60
Total Women	**10,670**	**12,144**	**14,955**	**16,741**	**19,564**	**21,650**
Under 30	9.17	7.02	6.27	4.09	4.14	3.91
30-34	26.21	25.20	24.16	24.81	21.26	18.65
35-39	24.35	25.07	25.20	23.64	24.40	26.82
40-44	14.80	17.42	18.88	20.92	20.68	21.37
45-49	7.84	8.66	9.80	11.05	13.33	15.13
50-54	7.41	6.42	5.88	5.67	7.17	7.40
55-59	5.73	6.00	5.71	5.47	4.52	3.32
60-64	2.75	2.65	2.75	3.18	3.02	2.29
65-69	1.33	1.09	1.18	0.84	1.01	0.92
70+	0.42	0.47	0.15	0.33	0.47	0.20
Median Age (years)	37.50	38.05	38.38	38.96	39.55	39.65

NOTE: "Biomedical science work force" consists of those working in a biomedical science field, including postdoctoral appointees, regardless of Ph.D. field. Estimates are subject to sampling error. Comparisons between 1991 estimates and those of earlier years should be made with caution due to changes in survey methodology.

SOURCE: NRC, Survey of Doctorate Recipients. (Biennial)

See Figures 3-1, 3-2, and 3-3.

APPENDIX F

APPENDIX TABLE F-4 Citizenship Status of Employed Biomedical Science Ph.D.s: 1981-1991 (in percent)

Citizenship Status	1981	1983	1985	1987	1989	1991
Total N	58,446	63,037	70,077	75,192	81,086	85,596
U.S. Citizen	95.0	95.8	95.7	95.5	95.3	94.0
U.S. Native born	*87.5*	*87.3*	*87.1*	*86.8*	*86.3*	*85.4*
Non-U.S. citizen	5.0	4.2	4.3	4.5	4.7	6.0
Total	100.0	100.0	100.0	100.0	100.0	100.0

NOTE: "Employed Biomedical Science Ph.D.s " are those with Ph.D.s in biomedical science fields, regardless of field of employment. Estimates are subject to sampling error. Comparisons between 1991 estimates and those of earlier years should be made with caution due to changes in survey methodology. Prior to 1991, the SDR collected data by mail methods only. In 1991, the survey had both a mail componenet and a telephone follow-up component. In this table, 1991 estimates are based on "mail-only" data to maintain greater comparability with earlier years.

SOURCE: NRC, Survey of Doctorate Recipients. (Biennial)

See Figure 3-4.

APPENDIX TABLE F-5 Total Employment by Sector of the U.S. Biomedical Science Work Force: 1981-1991

	Total	Academic	Industry	Government	Hospitals/Clinics	Other
Total Biomedical						
1981	63,942	40,855	11,323	6,084	2,983	2,697
1983	65,823	40,616	12,992	6,432	3,054	2,729
1985	74,217	45,142	13,263	6,835	3,636	3,341
1987	80,636	47,657	18,112	7,629	3,738	3,500
1989	88,344	51,134	21,658	8,257	3,674	3,621
1991	91,240	49,032	25,647	8,386	4,261	3,914
Men						
1981	53,354	33,447	10,277	5,062	2,542	2,026
1983	53,688	32,134	11,619	5,383	2,447	2,105
1985	59,469	35,247	13,366	5,472	2,877	2,507
1987	63,921	37,003	15,579	5,961	2,910	2,468
1989	68,820	38,878	18,180	6,461	2,795	2,506
1991	69,753	36,164	21,175	6,378	3,107	2,929
Women						
1981	10,588	7,408	1,046	1,022	441	671
1983	12,135	8,482	1,373	1,373	607	624
1985	14,748	9,895	1,897	1,363	759	834
1987	16,715	10,654	2,533	1,668	828	1,032
1989	19,524	12,256	3,478	1,796	879	1,115
1991	21,487	12,868	4,472	2,008	1,154	985

NOTE: "Biomedical science work force" consists of those working in a biomedical science field, including postdoctoral appointees, regardless of Ph.D. field. Cases with missing sector or work activity are excluded. See Appendix F-3. Estimates are subject to sampling error. Comparisons between 1991 estimates and those of earlier years should be made with caution due to changes in survey methodology.

SOURCE: NRC, Survey of Doctorate Recipients. (Biennial)

See Figure 3-5.

APPENDIX TABLE F-6 Doctorate Recipients in the Biomedical Sciences by Gender: 1981-1992

Year	Total	Gender (percent)	
		Male	Female
1981	3,444	70.5	29.5
1982	3,463	69.0	31.0
1983	3,317	65.1	34.9
1984	3,386	65.6	34.4
1985	3,197	64.0	36.0
1986	3,222	62.7	37.3
1987	3,200	61.8	38.2
1988	3,467	60.2	39.8
1989	3,490	59.2	40.8
1990	3,542	59.3	40.7
1991	3,745	58.6	41.4
1992	3,787	57.3	42.7

NOTE: Data limited to U.S. citizens and permanent residents. Cases with missing gender data are excluded. The Survey of Earned Doctorates is sponsored by five Federal Agencies: National Science Foundation (NSF); National Institutes of Health (NIH); U.S. Department of Education (USED); National Endowment for the Humanities (NIH); and the U.S. Department of Agriculture (USDA); and conducted by the National Research Council (NRC).

SOURCE: NRC, Survey of Earned Doctorates. (Annual)

See Figures 3-6 and 3-7.

APPENDIX F

APPENDIX TABLE F-7 Doctorate Recipients in the Biomedical Sciences by Citizenship: 1981-1992

Year	Total	Citizenship (percent)	
		U.S.	Non-U.S.
1981	3,745	87.7	12.3
1982	3,823	86.8	13.2
1983	3,679	86.3	13.7
1984	3,761	86.1	13.9
1985	3,637	83.9	16.1
1986	3,655	84.2	15.8
1987	3,737	80.8	19.2
1988	4,085	79.7	20.3
1989	4,174	78.7	21.3
1990	4,496	74.0	26.0
1991	4,881	71.3	28.7
1992	5,073	69.1	30.9

NOTE: Cases with missing data are excluded. The Survey of Earned Doctorates is sponsored by five Federal Agencies: National Science Foundation (NSF); National Institutes of Health (NIH); U.S. Department of Education (USED); National Endowment for the Humanities (NIH); and the U.S. Department of Agriculture (USDA); and conducted by the National Research Council (NRC).

SOURCE: NRC, Survey of Earned Doctorates. (Annual)

See Figure 3-8.

APPENDIX TABLE F-8 Unemployment Rates for Biomedical and Physical Science Ph.D.s and the U.S. Labor Force: 1973-1991

	1973	1975	1977	1979	1981	1983	1985	1987	1989	1991
Biomedical Sciences	1.1	1.2	1.6	1.2	1.2	1.4	1.1	1.1	0.9	1.5
Physical Sciences	1.6	1.2	1.1	1.0	0.6	1.0	0.6	1.1	0.6	1.4
U.S. Labor Force	4.9	8.5	7.1	5.8	7.6	9.6	7.2	6.2	5.3	6.7

NOTE: The unemployment rate is defined as the percent of the field-specific labor force who are not employed and are seeking work. The field-specific labor forces include individuals who have Ph.D.s in these fields and who are either employed or unemployed. Estimates are subject to sampling error. Comparisons between 1991 estimates and those of earlier years should be made with caution due to changes in survey methodology. Prior to 1991, the SDR collected data by mail methods only. In 1991, the survey had both a mail component and a telephone follow-up component. In this table, 1991 estimates are based on "mail-only" data to maintain greater comparability with earlier years.

SOURCES: NRC, Survey of Doctorate Recipients, (Biennial); Office of the President of the United States, *Economic Report of the President*, Washington, D.C.: U.S. Government Printing Office, 1993.

See Figure 3-9.

APPENDIX TABLE F-9 Underemployment Rates for Employed Biomedical and Physical Science Ph.D.s: 1973-1991

	1973	1975	1977	1979	1981	1983	1985	1987	1989	1991
Biomedical Sciences	1.0	0.8	0.9	0.9	0.9	1.0	0.9	1.1	0.8	1.2
Physical Sciences	1.5	1.2	1.0	0.8	0.6	0.9	0.5	0.7	0.8	1.1

NOTE: The underemployment rate is defined as the percent of the field-specific employed doctorates who are working part-time but seeking full-time jobs, or are working in a non-S&E job when an S&E job would be preferred. Estimates are subject to sampling error. Comparisons between 1991 estimates and those of earlier years should be made with caution due to changes in survey methodology. Prior to 1991, the SDR collected data by mail methods only. In 1991, the survey had both a mail component and a telephone follow-up component. In this table, 1991 estimates are based on "mail-only" data to maintain greater comparability with earlier years.

SOURCE: NRC, Survey of Doctorate Recipients. (Biennial)

See Figure 3-10.

APPENDIX TABLE F-10 Percentage of Biomedical Work Force at Career Age 4-5 who are in Postdoctoral Appointments: 1973-1991

Year	Percent
1973	6.6
1975	11.4
1977	13.5
1979	16.0
1981	20.7
1983	19.2
1985	20.9
1987	18.4
1989	21.4
1991	16.8

NOTE: "Biomedical work force" consists of those working in a biomedical science field, including postdoctoral appointees, regardless of Ph.D. field. Estimates are subject to sampling error. Comparisons between 1991 estimates and those of earlier years should be made with caution due to changes in survey methodology.

SOURCE: NRC, Survey of Doctorate Recipients. (Biennial)

See Figure 3-11.

APPENDIX TABLE F-11 Percentage of New Ph.D.s in the Biomedical and Physical Sciences Who Are U.S. Citizens or Permanent Residents with Definite Commitments: 1975-1992 [Index 1989=100]

Year	Biomedical Sciences	Physical Sciences
1975	96.6	95.2
1976	97.3	93.3
1977	95.2	94.2
1978	96.6	97.5
1979	99.9	100.4
1980	100	104.1
1981	100.2	105.2
1982	98.5	102.9
1983	98.8	99.6
1984	98.7	99.7
1985	98.9	100
1986	97.8	98.4
1987	97.3	98.7
1988	97.4	98.9
1989	100	100
1990	96.7	96.2
1991	95.6	92.1
1992	96.1	91.5

NOTE: See Appendix Table F-7. The raw percentages for 1989 are: 81.5 for the biomedical sciences and 78.9 for the physical sciences.

SOURCE: NRC, Survey of Earned Doctorates. (Annual)

See Figure 3-12.

APPENDIX TABLE F-12 Median Salaries of Employed Biomedical Science Ph.D.s Age 30-34 as a Percentage of Comparable Salaries for All Employed Ph.D. Scientists and Engineers Age 30-34: 1973-1991

Year	Percent
1973	94.2
1975	94.7
1977	95.6
1979	96.1
1981	93.1
1983	94.6
1985	90.5
1987	87.1
1989	88.3
1991	87.9

NOTE: Postdoctoral appointees are excluded from these calculations. Estimates are subject to sampling error. Comparisons between 1991 estimates and those of earlier years should be made with caution due to changes in survey methodology.

SOURCE: NRC, Survey of Doctorate Recipients. (Biennial)

See Figure 3-13.

APPENDIX TABLE F-13 Age Profile of the U.S. Behavioral Science Work Force by Gender: 1981-1991 (in percent)

Age	1981	1983	1985	1987	1989	1991
Total Behavioral	49,058	52,643	59,898	62,507	66,939	66,628
Under 30	5.21	3.74	2.51	1.58	1.46	1.49
30-34	19.86	18.36	16.81	13.33	10.44	9.28
35-39	23.40	23.46	21.98	22.93	21.73	18.32
40-44	14.32	16.56	20.15	21.26	21.61	24.10
45-49	10.72	11.44	11.77	13.99	16.77	19.33
50-54	10.93	9.50	9.73	9.50	9.98	10.99
55-59	8.58	9.09	8.75	8.68	7.98	7.45
60-64	4.48	5.33	5.57	5.52	6.50	5.73
65-69	1.88	1.89	2.04	2.55	2.55	2.54
70+	0.63	0.63	0.70	0.65	0.98	0.76
Median Age (years)	39.54	40.35	41.17	41.87	42.80	43.35
Total Men	35,722	37,223	40,968	41,523	43,175	40,430
Under 30	4.10	2.46	1.79	0.94	0.95	1.17
30-34	17.99	17.35	14.64	10.91	7.73	7.52
35-39	23.97	22.82	20.93	21.28	20.28	16.42
40-44	14.44	16.47	21.07	22.28	21.39	23.37
45-49	10.88	11.94	12.29	14.64	18.53	21.03
50-54	11.98	10.24	10.19	9.91	10.59	11.80
55-59	9.48	10.17	9.97	9.93	8.61	7.69
60-64	4.69	6.13	6.31	6.53	7.88	6.85
65-69	1.86	1.88	2.29	2.94	2.94	3.11
70+	0.61	0.55	0.53	0.63	1.10	1.05
Median Age (years)	40.86	41.74	42.50	43.28	44.42	44.86
Total Women	13,336	15,420	18,930	20,984	23,764	26,198
Under 30	8.19	6.84	4.07	2.85	2.39	2.00
30-34	24.87	20.78	21.48	18.13	15.36	12.00
35-39	21.88	25.02	24.26	26.18	24.36	21.25
40-44	14.01	16.76	18.15	19.23	22.01	25.24
45-49	10.28	10.23	10.66	12.71	13.59	16.72
50-54	8.11	7.72	8.75	8.68	8.87	9.75
55-59	6.15	6.50	6.10	6.22	6.82	7.07
60-64	3.91	3.40	3.96	3.53	4.00	4.02
65-69	1.92	1.93	1.49	1.78	1.85	1.65
70+	0.69	0.82	1.08	0.69	0.75	0.31
Median Age (years)	38.37	38.97	39.55	40.24	41.29	42.42

NOTES: "Behavioral science work force" consists of those working in a behavioral science field, including postdoctoral appointees, regardless of Ph.D. field. Estimates are subject to sampling error. Comparisons between 1991 estimates and those of earlier years should be made with caution due to changes in survey methodology.

SOURCE: NRC, Survey of Doctorate Recipients. (Biennial)

See Figures 4-1, 4-2, and 4-3.

APPENDIX TABLE F-14 Citizenship Status of Employed Behavioral Science Ph.D.s: 1981-1991 (in percent)

Citizenship Status	1981	1983	1985	1987	1989	1991
Total N	56,043	64,203	71,864	77,435	83,935	88,569
U.S. Citizen	98.0	97.9	98.0	98.1	98.0	97.5
U.S. Native born	*93.6*	*93.1*	*93.5*	*93.9*	*94.1*	*93.3*
Non-U.S. citizen	2.0	2.1	2.0	1.9	2.0	2.5
Total	100.0	100.0	100.0	100.0	100.0	100.0

NOTES: "Employed Behavioral Science Ph.D.s" are defined as those working in a behavioral science field, regardless of employment field. Estimates are subject to sampling error. Comparisons between 1991 estimates and those of earlier years should be made with caution due to changes in survey methodology. Prior to 1991, the SDR collected data by mail methods only. In 1991, the survey had both a mail component and a telephone follow-up component. In this table, 1991 estimates are based on "mail-only" data to maintain greater comparability with earlier years.

SOURCE: NRC, Survey of Doctorate Recipients. (Biennial)

See Figure 4-4.

APPENDIX TABLE F-15 Employment of the Behavioral Science Work Force by Sector: 1981-1991

	Total	Academic	Industry	Government	Hospitals/Clinics	Other
Total Behavioral						
1981	48,402	26,909	9,065	3,158	5,929	3,341
1983	52,282	27,859	11,941	3,108	6,021	3,353
1985	58,956	30,811	14,428	2,905	6,623	4,189
1987	62,288	30,914	15,904	3,717	7,343	4,410
1989	66,719	31,830	18,352	3,626	7,763	5,148
1991	65,949	27,936	20,764	3,884	8,529	4,836
Men						
1981	35,213	19,910	6,380	2,380	4,426	2,117
1983	37,031	20,189	8,100	2,348	4,437	1,957
1985	40,380	21,778	9,698	2,231	4,326	2,347
1987	41,401	21,425	9,959	2,611	4,845	2,561
1989	42,967	21,440	11,074	2,623	4,821	3,009
1991	40,025	17,641	11,937	2,677	5,238	2,532
Women						
1981	13,189	6,999	2,685	778	1,503	1,224
1983	15,251	7,670	3,841	760	1,584	1,396
1985	18,576	9,033	4,730	674	2,297	1,842
1987	20,887	9,489	5,945	1,106	2,498	1,849
1989	23,752	10,390	7,278	1,003	2,942	2,139
1991	25,924	10,295	8,827	1,207	3,291	2,304

NOTES: "Behavioral science work force" consists of those working in a behavioral science field, including postdoctoral appointees, regardless of Ph.D. field. Cases with missing sector or work activity are excluded. Estimates are subject to sampling error. Comparisons between 1991 estimates and those of earlier years should be made with caution due to changes in survey methodology.

SOURCE: NRC, Survey of Doctorate Recipients. (Biennial)

See Figure 4-5.

APPENDIX TABLE F-16 Doctorate Recipients in the Behavioral Sciences by Gender: 1981-1992

Year	Total	Gender (percent)	
		Male	Female
1981	4,149	55.8	44.2
1982	3,836	54.6	45.4
1983	3,989	51.8	48.2
1984	3,807	49.1	50.9
1985	3,656	48.8	51.2
1986	3,631	47.9	52.1
1987	3,540	46.0	54.0
1988	3,434	44.5	55.5
1989	3,399	43.1	56.9
1990	3,668	41.1	58.9
1991	3,741	38.0	62.0
1992	3,647	41.3	58.7

NOTE: Data limited to U.S. citizens and permanent residents. Cases with missing gender data are excluded. The Survey of Earned Doctorates is sponsored by five Federal Agencies: National Science Foundation (NSF); National Institutes of Health (NIH); U.S. Department of Education (USED); National Endowment for the Humanities (NIH); and the U.S. Department of Agriculture (USDA); and conducted by the National Research Council (NRC).

SOURCE: NRC, Survey of Earned Doctorates. (Annual)

See Figures 4-6 and 4-7.

APPENDIX TABLE F-17 Doctorate Recipients in the Behavioral Sciences by Citizenship: 1981-1992

Year	Total	Citizenship (percent)	
		U.S.	Non-U.S.
1981	4,323	94.0	6.0
1982	3,988	94.1	5.9
1983	4,159	93.7	6.3
1984	3,989	93.6	6.4
1985	3,823	93.2	6.8
1986	3,812	92.4	7.6
1987	3,708	92.5	7.5
1988	3,620	92.0	8.0
1989	3,638	90.8	9.2
1990	3,923	90.7	9.3
1991	4,026	89.9	10.1
1992	3,966	88.7	11.3

NOTE: Cases with missing data are excluded. The Survey of Earned Doctorates is sponsored by five Federal Agencies: National Science Foundation (NSF); National Institutes of Health (NIH); U.S. Department of Education (USED); National Endowment for the Humanities (NIH); and the U.S. Department of Agriculture (USDA); and conducted by the National Research Council (NRC).

SOURCE: NRC, Survey of Earned Doctorates. (Annual).

See Figure 4-8.

APPENDIX F

APPENDIX TABLE F-18 Unemployment Rates for Behavioral and Physical Science Ph.D.s and the U.S. Labor Force: 1973-1991

	1973	1975	1977	1979	1981	1983	1985	1987	1989	1991
Behavioral Sciences	1.2	1.0	1.6	1.3	1.2	1.3	1.0	1.0	1.2	1.4
Physical Sciences	1.6	1.2	1.1	1.0	0.6	1.0	0.6	1.1	0.6	1.4
U.S. Labor Force	4.9	8.5	7.1	5.8	7.6	9.6	7.2	6.2	5.3	6.7

NOTE: See Appendix Table F-8. Comparisons between 1991 estimates and those of earlier years should be made with caution due to changes in survey methodology. Prior to 1991, the SDR collected data by mail methods only. In 1991, the survey had both a mail component and a telephone follow-up component. In this table, 1991 estimates are based on "mail-only" data to maintain greater comparability with earlier years.

SOURCES: NRC, Survey of Doctorate Recipients, (Biennial); Office of the President of the United States (1993), *Economic Report of the President*, Washington, D.C.: U.S. Government Printing Office.

See Figure 4-9.

APPENDIX TABLE F-19 Underemployment Rates for Employed Behavioral and Physical Science Ph.D.s : 1973-1991

	1973	1975	1977	1979	1981	1983	1985	1987	1989	1991
Behavioral	1.1	1.1	2.1	2.1	2.0	2.5	2.4	2.5	2.5	2.6
Physical Sciences	1.5	1.2	1.0	0.8	0.6	0.9	0.5	0.7	0.8	1.1

NOTE: See Appendix Table F-9. Comparisons between 1991 estimates and those of earlier years should be made with caution due to changes in survey methodology. Prior to 1991, the SDR collected data by mail methods only. In 1991, the survey had both a mail component and a telephone follow-up component. In this table, 1991 estimates are based on "mail-only" data to maintain greater comparability with earlier years.

SOURCE: NRC, Survey of Doctorate Recipients. (Biennial)

See Figure 4-10.

APPENDIX TABLE F-20 Percentage of New Ph.D.s Who Are U.S. Citizens or Permanent Residents with Definite Commitments in the Behavioral and Physical Sciences: 1975-1992. [Index 1989=100]

Year	Behavioral Sciences	Physical Sciences
1975	102.5	95.2
1976	99.2	93.3
1977	97	94.2
1978	94.5	97.5
1979	96.7	100.4
1980	99.7	104.1
1981	100.1	105.2
1982	95.6	102.9
1983	94.2	99.6
1984	94.2	99.7
1985	96.6	100
1986	98.8	98.4
1987	95	98.7
1988	98.8	98.9
1989	100	100
1990	97.4	96.2
1991	97.7	92.1
1992	96.1	91.5

NOTE: See Appendix Table F-7. The raw percentages for 1989 are: 69.8 for the behavioral sciences and for the 78.9 physical sciences.

SOURCE: NRC, Survey of Earned Doctorates. (Annual)

See Figure 4-11.

APPENDIX TABLE F-21 Median Salaries of Employed Behavioral Science Ph.D.s Age 30-34 as a Percentage of Comparable Salaries for All Employed Scientists and Engineers Age 30-34: 1973-1991

Year	Percent
1973	95.9
1975	95.2
1977	92.6
1979	92.1
1981	88.4
1983	87.3
1985	85.2
1987	85.6
1989	86.2
1991	86.0

NOTE: See Appendix Table F-12. Comparisons between 1991 estimates and those of earlier years should be made with caution due to changes in survey methodology.

SOURCE: NRC, Survey of Doctorate Recipients. (Biennial)

See Figure 4-12.

APPENDIX F

APPENDIX TABLE F-22 Trends in the Supply of and Demand for Clinical Scientists

	Fiscal Year				
	1981	1982	1983[a]	1984	1985
SUPPLY INDICATORS (New Entrants):					
Professional doctorates participating in NIH training grants and fellowships[b]	2,172	2,153	2,290	2,367	2,385
M.D. degrees awarded	15,673	15,985	15,801	16,369	16,321
DEMAND INDICATORS:					
Expenditures for clinical R & D in medical schools (1987 $, mil.)	$783	$773	$811	$856	$961
Professional service income in medical schools (1987 $, mil.)	$1,301	$1,510	$1,829	$1,979	$2,340
Total revenue (all sources), (1987 $, mil.)	8,143	8,611	9,380	9,901	10,676
Budgeted vacancies in medical schools:					
Clinical departments	*2,231*	*2,264*	*2,270*	*2,402*	*2,572*
Basic science departments	*656*	*668*	*671*	*705*	*801*
Clinical faculty/student ratio[c]	0.331	0.341	0.346	0.349	0.354
LABOR FORCE					
M.D.s primarily engaged in research[d]	17,901	16,743	18,535	22,945	23,268
Full-time faculty in clinical departments	37,716	40,148	41,938	43,443	45,007
NIH research grants awarded to M.D.s:[e]					
Number of competing grants	*1,868*	*1,791*	*1,962*	*2,151*	*2,229*
% of total competing grants	*34%*	*32%*	*32%*	*34%*	*31%*
M.D. applicants for NIH research grants:					
Number of competing applicants	*4,525*	*4,716*	*4,658*	*4,844*	*5,264*
% of total competing applicants	*31%*	*30%*	*30%*	*30%*	*30%*
M.D. success rate:					
M.D. awards/M.D. applicants	*41%*	*38%*	*42%*	*44%*	*42%*
M.D. awardees/all awardees	*34%*	*32%*	*32%*	*33%*	*31%*
ENROLLMENTS:					
Medical students	65,412	66,484	66,886	67,437	67,086
Residents and clinical fellows[f]	52,871	57,504	59,138	60,442	63,507
Total	118,283	123,988	126,024	127,879	130,593

a Financial data from the University of Washington and Mayo Medical School were included for the first time in 1983.
b Includes Fogarty International Center programs.
c Ratio of full-time clinical faculty to a 4-year weighted average of total enrollments of medical students, residents and clinical fellows(WS), where $(WS)_t = 1/6(S_t + 2S(t-1) + 2S(t-2) + S(t-3))$.

APPENDIX F

	1986	1987	1988	1989	1990	1991	Annual Growth Rate	Latest Annual Change	Average Annual Change
	2,599	2,619	2,631	2,593	2,437	2,232	3.2%	-1.4%	69
	15,979	15,719	15,932	15,617	15,338	15,435	0.2%	-1.8%	34
	$1,061	$1,118	$1,204	$1,338	$1,516	n/a	6.8%	11.1%	$65
	$2,422	$2,745	$3,038	$4,333	$4,958	$5,634	14.5%	14.4%	$349
	11,451	12,589	13,577	15,792	16,626	n/a	7.7%	16.3%	$830
	2,345	*2,510*	*2,564*	*2,856*	*3,136*	*3,601*	*3.7%*	*9.8%*	*94*
	757	*785*	*718*	*720*	*667*	*643*	*-0.7%*	*-7.4%*	*-5*
	0.368	0.377	0.408	0.432	0.442	0.452	2.9%	2.2%	0.011
	17,847	n/a	16,586	16,941	16,930	n/a	1.7%	2.1%	270
	47,193	48,834	52,260	55,468	57,202	59,189	4.8%	3.1%	2,104
	2,075	*2,394*	*2,143*	*1,907*	*1,860*	*2,141*	*-1.3%*	*-11.0%*	*-26*
	30%	*32%*	*30%*	*31%*	*32%*	*31%*	*-0.7%*	*3.0%*	*-0.23%*
	5,359	*5,262*	*5,290*	*5,311*	*5,422*	*5,486*	*1.9%*	*0.4%*	*90*
	29%	*29%*	*28%*	*28%*	*28%*	*29%*	*-1.4%*	*-1.5%*	*-0.41%*
	39%	*45%*	*41%*	*36%*	*34%*	*39%*	*-3.1%*	*-11.4%*	*-1.32%*
	30%	*32%*	*30%*	*31%*	*32%*	*31%*	*-0.6%*	*-0.9%*	*-0.20%*
	66,607	65,918	65,711	65,150	65,081	64,996	0.4%	-0.1%	227
	60,875	61,539	62,880	64,867	67,110	68,329	2.7%	3.5%	1,538
	127,482	127,457	128,591	130,017	132,191	133,325	1.4%	1.3%	1,766

d There is some question about the interpretation of these data since they include many trainee graduate medical programs. 1984 data are previously unpublished; AMA reports them to be to "within 100 persons". 1987 data remain unpublished.

e Data for M.D.s do NOT exclude those who may also have a Ph.D. degree. As a result, the percent M.D. awardees and percent Ph.D. awardees, can be greater than 100%.

f The residents and clinical fellows reported here include only those in accredited programs af with medical schools.

SOURCE: National Institutes of Health, special tabulations, 1992.

APPENDIX TABLE F-23 R&D Expenditures in U.S. Medical Schools, by Type of Institutional Control, 1964-91, ($ thousands)

Fiscal Year	Total R&D Expenditures					
	Current Dollars [a]			1987 Dollars		
	Total	Public	Private	Total	Public	Private
1964	310,412	128,710	181,702	1,120,621	464,657	655,964
1965	344,787	143,627	201,160	1,214,039	505,729	708,310
1966	377,028	155,960	221,068	1,327,563	549,155	778,408
1967	422,467	178,881	243,586	1,436,963	608,439	828,524
1968	470,958	202,440	268,518	1,554,317	668,119	886,198
1969	489,300	196,800	292,500	1,469,369	590,991	878,378
1970	498,066	205,962	292,104	1,418,991	586,786	832,205
1971	499,841	207,346	292,495	1,424,048	590,729	833,319
1972	558,120	227,638	330,482	1,508,432	615,238	893,195
1973	606,921	264,808	342,113	1,564,229	682,495	881,735
1974	657,287	300,479	356,808	1,463,891	669,218	794,673
1975	784,537	363,893	420,644	1,594,587	739,620	854,967
1976	839,170	385,857	453,313	1,705,630	784,262	921,368
1977	973,827	449,709	524,118	1,862,002	859,864	1,002,138
1978	1,046,121	490,029	556,092	1,871,415	876,617	994,798
1979	1,190,689	585,488	605,201	1,817,846	893,875	923,971
1980	1,352,409	677,085	675,324	1,886,205	944,331	941,874
1981	1,477,919	766,565	711,354	2,061,254	1,069,128	992,126
1982	1,605,585	828,954	776,631	2,034,962	1,050,639	984,323
1983[c]	1,787,532	937,510	850,022	2,133,093	1,118,747	1,014,346
1984	1,997,317	1,065,607	931,710	2,194,854	1,170,997	1,023,857
1985	2,301,421	1,229,256	1,072,165	2,437,946	1,302,178	1,135,768
1986	2,554,561	1,400,225	1,154,336	2,706,103	1,483,289	1,222,814
1987	2,800,837	1,500,175	1,300,662	2,890,441	1,548,168	1,342,272
1988	3,043,336	1,592,243	1,451,093	3,043,336	1,592,243	1,451,093
1989	3,633,314	1,992,784	1,640,529	3,351,765	1,838,362	1,513,403
1990	4,264,303	2,133,545	2,130,757	3,777,061	1,889,765	1,887,296
1991	4,640,646	2,304,664	2,335,982	4,110,404	2,041,332	2,069,072

a Figures were obtained from the Association of American Medical Colleges (1972-92a, annual tabulations). Because AAMC data were incomplete, figures for 1963 and 1969 were obtained from the American Medical Association (1960-90). Items may not sum to totals due to rounding.
b From the U.S. Bureau of the Census (1992).
c Estimates for 1964-1975 were derived from data supplied by John James, NIH Division of Research Grants; 1976-1984 data were estimated by the NRC Committee on National Needs for Biomedical and Behavioral Research Personnel; 1985-1990 were derived from the NSF Survey of Federal Support to Universities.
d Financial data from the University of Washington and Mayo Medical School were included for the first time in 1983.
SOURCE: National Institutes of Health, special tabulations, 1992.

APPENDIX F

Implicit GNP Price Deflator [b] (1987 = 100)	NIH Clinical Research as a % of Total NIH Research Obligation [c]	Estimated Clinical R & D [d] 1987 Dollars		
		Total	Public	Private
27.7	15.0	168,093	69,699	98,395
28.4	16.5	200,316	83,445	116,871
29.4	18.0	238,961	98,848	140,114
30.3	20.0	287,393	121,688	165,705
31.7	22.5	349,721	150,327	199,395
33.3	25.0	367,342	147,748	219,595
35.1	28.0	397,318	164,300	233,017
37	30.0	427,215	177,219	249,996
38.8	32.0	482,698	196,876	285,822
41.3	34.0	531,838	232,048	299,790
44.9	34.0	497,723	227,534	270,189
49.2	39.0	621,889	288,452	333,437
52.3	37.0	631,083	290,177	340,906
55.9	39.0	726,181	335,347	390,834
60.3	41.0	767,280	359,413	407,867
65.5	38.0	690,781	339,672	351,109
71.7	39.0	735,620	368,289	367,331
78.9	38.0	783,276	406,269	377,008
83.8	38.0	773,286	399,243	374,043
87.2	38.0	810,575	425,124	385,452
91	39.0	855,993	456,689	399,304
94.4	39.4	961,200	513,405	447,795
96.9	39.2	1,061,358	581,759	479,598
100.0	38.7	1,117,620	598,616	519,004
103.9	39.6	1,203,873	629,855	574,019
108.4	39.9	1,337,651	733,669	603,982
112.9	40.1	1,515,779	758,385	757,394
117.0	n/a	n/a	n/a	n/a

APPENDIX TABLE F-24 Average Clinical R and D Expenditures and Professional Service Income per U.S. Medical School, by Type of Institutional Control, 1966-90 (1987 $ thousands)

| | Clinical R and Expenditures | | | | | [a] |
| | Average per School | | | Number of Schools Reporting | | |
Fiscal Year	Total	Public	Private	Total	Public	Private
1966	2,685	2,103	3,336	89	47	42
1967	2,994	2,296	3,854	96	53	43
1968	3,605	2,733	4,747	97	55	42
1969	4,037	2,841	5,631	91	52	39
1970	3,973	2,833	5,548	100	58	42
1971	4,188	2,954	5,952	102	60	42
1972	4,686	3,337	6,496	103	59	44
1973	5,163	3,867	6,972	103	60	43
1974	4,484	3,447	6,004	111	66	45
1975	5,603	4,305	7,578	111	67	44
1976	5,635	4,331	7,576	112	67	45
1977	6,260	4,791	8,496	116	70	46
1978	6,394	4,992	8,497	120	72	48
1979	5,709	4,653	7,315	121	73	48
1980	6,080	4,977	7,816	121	74	47
1981	6,527	5,565	8,021	120	73	47
1982	6,287	5,395	7,634	123	74	49
1983[b]	6,590	5,745	7,866	123	74	49
1984	6,903	6,171	7,986	124	74	50
1985	7,569	6,845	8,611	127	75	52
1986	8,357	7,757	9,223	127	75	52
1987	8,800	8,089	9,793	127	74	53
1988	9,479	8,512	11,039	127	74	52
1989	10,616	9,914	11,615	126	74	52
1990	12,030	10,248	14,565	126	74	52

a From the Association of American Medical Colleges (1972-92, special tabulations generated annually from 1982-92).

b Financial data from the University of Washington and Mayo Medical School were included for the first time in 1983.

SOURCE: National Institutes of Health, special tabulations, 1992.

APPENDIX F

Professional Service Income						Sum of Average Clinical R&D plus Professional Service Income per School		
Average per School			Number of Schools Reporting [a]					
Total	Public	Private	Total	Public	Private	Total	Public	Private
2,521	2,451	2,604	33	18	15	5,206	4,554	5,940
2,322	2,256	2,405	43	24	19	5,316	4,552	6,258
2,670	2,686	2,648	56	33	23	6,276	5,419	7,396
3,440	3,422	3,466	57	33	24	7,477	6,263	9,097
3,614	3,543	3,716	71	42	29	7,587	6,376	9,264
4,474	4,298	4,755	70	43	27	8,663	7,252	10,707
5,017	4,744	5,389	71	41	30	9,703	8,081	11,885
5,499	5,449	5,563	70	39	31	10,663	9,316	12,535
6,152	5,899	6,583	73	46	27	10,636	9,347	12,587
7,956	7,148	9,250	78	48	30	13,559	11,453	16,828
9,330	8,371	10,740	84	50	34	14,964	12,702	18,315
11,129	8,745	14,807	89	54	35	17,389	13,535	23,303
11,002	8,772	14,376	93	56	37	17,396	13,764	22,874
11,975	9,672	15,621	93	57	36	17,684	14,325	22,935
12,790	10,148	17,192	96	60	36	18,869	15,125	25,008
13,008	10,229	17,540	100	62	38	19,535	15,795	25,562
14,801	11,500	19,918	102	62	40	21,088	16,895	27,552
17,096	13,402	23,043	107	66	41	23,686	19,147	30,909
18,496	14,465	24,985	107	66	41	25,399	20,637	32,971
21,269	15,649	29,700	110	66	44	28,838	22,494	38,311
21,820	17,580	28,525	111	68	43	30,177	25,337	37,748
24,289	18,732	32,687	113	68	45	33,089	26,821	42,479
26,886	21,490	35,039	113	68	45	36,365	30,002	46,078
39,038	30,768	52,115	111	68	43	49,654	40,683	63,730
43,877	35,611	56,839	113	69	44	55,907	45,859	71,404

APPENDIX TABLE F-25 First Time R01 Applicants by Degree and Prior Research Training Experience

FY of First Application	Total Applicants			
	M.D.s	Ph.D.s	M.D.-Ph.D.s	Others
1964	1,087	1,453	103	442
1965	1,035	1,686	107	204
1966	1,081	1,613	81	244
1967	987	1,530	113	216
1968	826	1,582	96	198
1969	691	1,595	120	158
1970	600	1,766	83	145
1971	569	1,707	98	154
1972	654	1,820	111	139
1973	675	1,897	148	146
1974	697	2,005	123	137
1975	634	1,908	138	136
1976	845	2,288	167	168
1977	896	2,509	142	200
1978	907	2,403	146	168
1979	855	2,556	133	168
1980	906	2,501	133	194
1981	796	2,245	127	312
1982	682	2,168	64	177
1983	699	1,999	64	228
1984	700	2,070	114	158
1985	727	2,123	144	120
1986	669	2,175	157	131
1987	620	1,532	108	115
1988	696	1,913	139	105
1989	711	2,019	184	87

SOURCE: National Institutes of Health, special tabulations, 1991.

APPENDIX F

Applicants with Research Training				Percent With Research Training			
M.D.s	Ph.D.s	M.D.-Ph.D.s	Others	M.D.s	Ph.D.s	M.D.-Ph.D.s	Others
386	458	35	103	35.51	31.52	33.98	23.30
409	600	54	36	39.52	35.59	50.47	17.65
498	640	45	38	46.07	39.68	55.56	15.57
446	651	54	28	45.19	42.55	47.79	12.96
412	733	48	25	49.88	46.33	50.00	12.63
373	738	67	20	53.98	46.27	55.83	12.66
351	928	47	23	58.50	52.55	56.63	15.86
356	893	64	23	62.57	52.31	65.31	14.94
398	945	66	17	60.86	51.92	59.46	12.23
399	1,031	91	24	59.11	54.35	61.49	16.44
421	1,127	69	10	60.40	56.21	56.10	7.30
386	1,078	91	18	60.88	56.50	65.94	13.24
488	1,288	104	23	57.75	56.29	62.28	13.69
556	1,449	82	19	62.05	57.75	57.75	9.50
536	1,395	95	36	59.10	58.05	65.07	21.43
509	1,402	89	42	59.53	54.85	66.92	25.00
550	1,391	93	31	60.71	55.62	69.92	15.98
431	1,288	77	107	54.15	57.37	60.63	34.29
418	1,270	43	48	61.29	58.58	67.19	27.12
394	1,152	40	74	56.37	57.63	62.50	32.46
396	1,196	76	36	56.57	57.78	66.67	22.78
385	1,226	94	21	52.96	57.75	65.28	17.50
335	1,205	93	22	50.07	55.40	59.24	16.79
308	800	72	21	49.68	52.22	66.67	18.26
311	930	79	19	44.68	48.61	56.83	18.10
298	954	107	16	41.91	47.25	58.15	18.39

APPENDIX TABLE F-26 First Time Grant* Recipients Degree and Prior Research Training Experience

FY of First Application	Total Recipients			
	M.D.s	Ph.D.s	M.D.-Ph.D.s	Others
1963	100	100	10	1,500
1964	652	867	75	137
1965	582	688	74	85
1966	616	660	58	82
1967	542	728	61	74
1968	363	527	59	40
1969	376	625	70	32
1970	266	483	46	22
1971	310	581	64	12
1972	432	803	85	23
1973	314	601	63	19
1974	460	1,087	97	17
1975	444	1,181	121	28
1976	406	917	93	27
1977	419	918	80	24
1978	481	1,247	109	23
1979	556	1,516	97	40
1980	469	1,169	82	34
1981	464	1,121	95	77
1982	409	1,109	53	53
1983	494	1,283	48	86
1984	546	1,264	90	94
1985	592	1,385	122	140
1986	521	1,467	131	109
1987	577	1,388	110	79
1988	561	1,453	128	88
1989	512	1,251	146	94

*K,R,U,M,P Awards
SOURCE: National Institutes of Health, special tabulations, 1991.

APPENDIX F

Recipients with Research Training				Percent With Research Training			
M.D.s	Ph.D.s	M.D.-Ph.D.s	Others	M.D.s	Ph.D.s	M.D.-Ph.D.s	Others
35	38	3	586	35.00	38.00	30.00	39.07
265	345	31	37	40.64	39.79	41.33	27.01
279	321	33	21	47.94	46.66	44.59	24.71
328	328	29	18	53.25	49.70	50.00	21.95
296	379	40	17	54.61	52.06	65.57	22.97
224	321	32	9	61.71	60.91	54.24	22.50
258	394	40	10	68.62	63.04	57.14	31.25
188	337	30	6	70.68	69.77	65.22	27.27
215	406	47	4	69.35	69.88	73.44	33.33
303	552	55	5	70.14	68.74	64.71	21.74
186	406	45	2	59.24	67.55	71.43	10.53
294	725	62	2	63.91	66.70	63.92	11.76
294	835	82	4	66.22	70.70	67.77	14.29
251	629	60	3	61.82	68.59	64.52	11.11
279	647	50	2	66.59	70.48	62.50	8.33
332	852	63	4	69.02	68.32	57.80	17.39
401	1,037	72	12	72.12	68.40	74.23	30.00
336	801	64	9	71.64	68.52	78.05	26.47
317	765	65	48	68.32	68.24	68.42	62.34
287	804	36	21	70.17	72.50	67.92	39.62
303	891	31	27	61.34	69.45	64.58	31.40
343	854	67	22	62.82	67.56	74.44	23.40
363	922	83	47	61.32	66.57	68.03	33.57
289	892	82	27	55.47	60.80	62.60	24.77
317	842	79	11	54.94	60.66	71.82	13.92
322	862	82	10	57.40	59.33	64.06	11.36
269	689	99	13	52.54	55.08	67.81	13.83

APPENDIX G

MULTISTATE LIFE TABLE METHODOLOGY AND PROJECTIONS

Multistate period life tables were used in this report to develop projections of numbers of new Ph.D.s that would be needed in the future to sustain certain growth rates of the labor force. This Appendix describes briefly the method used to generate these results.[1]

Life table techniques are sometimes the only way to obtain estimates of certain statistics describing mobility and career characteristics of a population, especially those related to rates of occurrence of events, duration of time spent in an activity, and rates of attrition or exit from a population (due to death, retirement, job changing, etc.) even when we have not observed the full lifetimes of the scientists with our data (which is often the case with most data sets).[2] Moreover, life table methods provide a useful way of organizing various age-specific rates (rates of entering the labor force, changing jobs, moving abroad, retiring, and dying) into a logical framework, which can then be used to make projections of various characteristics of a population, such as its age distribution.

Multistate life tables, an extension of basic life tables, allow greater complexity to enter the analysis: people can enter as well as exit a population and can move back and forth across a variety of states within a population. Life tables may be classified as period versus cohort life tables. The Panel on Estimation Procedures decided that it was more practical, for the purposes of making projections, to use the former.

Construction of period life tables involves taking age-specific transition rates (job changing, unemployment, retirement, and death rates) prevailing during a particular period (e.g., 1989-1991) and applying them to a hypothetical (synthetic) cohort of people (actually a "synthetic" cohort of new Ph.D.s). Probabilities are then calculated of entrance or exit from a state (probability of entering a postdoctoral position, for example) and length of time spent in various states, as implied by the life table. Statistics of interest were developed for three time periods (1985-1991, 1979-1985, 1973-1979). Period life table results are conventionally referred to as "expected" quantities (i.e., "expected fraction of people who...,", "expected length of time..") because of the nature of the methodology: constructing a single hypothetical Ph.D. cohort that experiences the current transition rates taken from a variety of Ph.D. cohorts.

The data for the life table analysis come from the longitudinal Survey of Doctorate Recipients (SDR), a sample survey that follows a group of Ph.D.s over time, interviewing basically the same people every 2 years. A general description follows (for details, see NRC, 1991 SDR Methodological Report, forthcoming). In 1973 an initial sample of science or engineering Ph.D.s living in the United States was drawn, and those sample members have been followed through time. New Ph.D.s enter the SDR in 1975 and each subsequent SDR year (1977, 1979,...1991) and are followed over time as well. Individuals are followed until they reach a certain cutoff point that depends on the survey year at which they entered [typically 42 years after the Ph.D., although in recent SDR waves, they are followed until they reach age 70 or until they drop out for other reasons (nonresponse or death)].

The form of the data on which the life tables are based consists mostly of large sets of transition tables constructed from the SDR by National Research Council (NRC) staff (but death rates are obtained from TIAA-CREF data from the late 1980s, taken from Bowen and Sosa, 1991). For every pair of biennial survey-interview years ("waves") in the SDR (1973-1975 as the first pair, 1989-1991 as the last pair), the number of people moving between various states within those 2 years were obtained. These states were: postdoctorate, R&D employment[3] within one's broad Ph.D.

field (biomed, behavioral, other), non-R&D employment within broad Ph.D. field, employment outside of broad Ph.D. field, out of labor force or unemployed (combined), leaving the country, retirement, and death. All of the biennial transition proportions were obtained by 2 year age group by broad Ph.D. field, and by sex. The survey observations (i.e., people) in one set of biennial transitions are often the same people in subsequent sets (though older).[4]

Another data ingredient for the life tables is the distribution of states (same as above) of "new entrants to the SDR" for SDR waves 1975-1991, again by age, sex, and Ph.D. field. These are used as estimates of numbers of new Ph.D.s in each survey year.

These transition data sets constructed by NRC staff were transformed into proportions to be used as input into a Multistate Life Table program (Tiemeyer and Ulmer, 1991). Initial work involved explorations of data quality, sample sizes, and the stability of rates over time. To have large enough sample sizes for (what we would hope to be) reliable estimates of sex differences in career patterns as well estimates of how the career patterns have changed over time, it was necessary to aggregate the data into three broad time periods (as opposed to looking at a larger number of time periods): 1985-1991, 1979-1985, 1973-1979.

Projection Models

Life table construction begins by calculating a matrix containing the proportion of individuals exiting an origin state for each possible destination state between ages x and x+2 (in our case). This matrix is called M_x.

Our projection models hold the population of those employed "in field" to some constant growth rate. The following algorithm is used:

1. Survive the current specified Ph.D. population forward 2 years.
2. Calculate the number of individuals employed "in field".
3. Calculate the differences between the target "in field" population and the number "in field" in the survived current population. This yields the number of new entrants needed to increase the "in field" population to its target size.
4. Divide the result of (3) by the proportion of new entrants who enter an "in field" employment state on receiving their Ph.D..
5. Use the result of (4) as the number of new entrants who would have had to enter the population between year y and y+2 to attain the target "in field" population. Add these individuals into the life table, distributed approximately by age and destination state.

Let $N_{x,y}$ represent the number of individuals in the specified Ph.D. population in each employment state at age X for a given year Y. $N_{x,y}$ is a k by k matrix, where k equals the number of states in the model. The columns indicate origin states (in the base year) and the rows destination states. So $N_{x,1995}[4,1]$ would equal the number of people who were in the 4th state (out of field employment) in 1995 who were in the 1st state (in field post-doc) in 1991. For the base year, the off-diagonal elements of $N_{x,1991}$ are all 0 and the on-diagonal elements are equal the number of individuals age X in 1991 in the specified Ph.D. population in each employment state.

Let $N^-_{x,y}$ represent the number of individuals in each state (by origin state in 1991) in year y, BEFORE new entrants between year y and y-2 are added into the life table. Then $N^-_{x,y}$ is given by:

$$N^-_{x,y} = N_{x,y-2} \cdot \left(I + \frac{M_x}{2}\right)^{-1} \cdot \left(I - \frac{M_x}{2}\right)$$

Let F_{1991} represent the total number of individuals in the specified Ph.D. population employed "in field" in year 1991. Then F_{1991} is given by:

$$F_{1991} = \sum_{x=25}^{71} \sum_{o=1}^{3} \sum_{d=1}^{8} N_{x,1991}[d,o]$$

where x represents age, o represents origin of state, d represents destination state, $N_{x,1991}[o,d]$ represents the dth row and the oth column of $N_{x,1991}$, and where states 1 through 3 represent the employed "in field" states.

Let F^-_y represent the total number of individuals in the specified Ph.D. population employed "in field" in year Y who were in the specified Ph.D. population (although not necessarily employed "in field") in year Y-2. Then F^-_y is given by:

$$F^-_y = \sum_{x=25}^{71} \sum_{o=1}^{3} \sum_{d=1}^{8} N^-_{x,y}[d,o]$$

Let G represent the assumed 2-year growth rate for F_y. Then the target employed "in field" population size for any given year is:

Target "In Field" Population Size (y) = $F_{1991} \cdot (1+G)^{y-1991}$

Let D_x represent the proportionate distribution by age and state of new entrants to the specified Ph.D. population over the two year period between Y and Y+2. D_x is a 1 by k vector with each column representing the proportion of all new entrants who are age X who enter that state on receiving their Ph.D. Summing D_x across all ages and states should equal one.

Finally, let R represent the proportion of all new entrants who enter an "in field" state on receiving their Ph.D. Then R is given by:

$$R = \sum_{x=25}^{71} \sum_{d=1}^{3} D_x[d]$$

Given $N_{x,1991}$, M_x, D_x, and G, then $N_{x,y}$ can be calculated for any y greater than 1991 (in increments of 2-years) by iterating through the formula:

$$N_{x,y|y1991} = N^-_{x,y} + \frac{\left(\left(F_{1991} \cdot (1+G)^{y-1991}\right) - F^-_y\right)}{R} \cdot ((S \cdot D_x) \times I)$$

which expands to:

$$N_{x,y|y>1991} = N_{x,y-2} \cdot \left(I + \frac{M_x}{2}\right)^{-1} \cdot \left(I - \frac{M_x}{2}\right) +$$

$$\frac{\left(F_{1991} \cdot (1+G)^{y-1991}\right) - \sum_{x=25}^{71}\sum_{o=1}^{3}\sum_{d=1}^{8}\left(N_{x,y-2} \cdot \left(I + \frac{M_x}{2}\right)^{-1} \cdot \left(I - \frac{M_x}{2}\right)\right)[d,o]}{R}$$

$$\cdot ((S \cdot D_x) \times I)$$

where I is a k by k identity matrix, S is a k by 1 vector of ones, and the symbol x designates an element-wise matrix multiplication operation. (The operation ((S • D) x I) merely takes the D_x vector and turns it into a matrix with the elements of D_x on the main diagonal and zeros on the off diagonals).

The total number of new entrants is given by the component:

$$\frac{\left(F_{1991} \cdot (1+G)^{y-1991}\right) - \sum_{x=25}^{71}\sum_{o=1}^{3}\sum_{d=1}^{8}\left(N_{x,y-2} \cdot \left(I + \frac{M_x}{2}\right)^{-1} \cdot \left(I - \frac{M_x}{2}\right)\right)[d,o]}{R}$$

$$\cdot ((S \cdot D_x) \times I)$$

Projections were made separately for each of 4 populations:

- biomedical Ph.D.'s in biomedical employment fields,
- non-biomedical Ph.D.'s in biomedical employment fields,
- behavioral Ph.D.'s in behavioral employment fields, and
- non-behavioral Ph.D.'s in behavioral employment fields.

To illustrate the use of life table analysis in generating projections of workforce variables, the Panel, as an exploratory exercise, chose to generate estimates of job openings. Given the uncertainty associated with efforts to project demand, the Panel examined three growth rate scenarios based on the average annual growth in the biomedical and behavioral science workforces between 1981 and 1991: zero growth; one-half the 1981-1991 average annual growth; and the average annual growth.[5] Estimates of "net separations"[6] were generated using the life tables. Estimates of needed job openings for the alternative growth scenarios were derived by adding to these separations the number of additional job openings that would need to be created to attain the particular target rate of growth.

In generating the estimates of job openings, the following assumptions were made:

1. There is never a negative number of new entrants. If there is a surplus in the employed "in field" population at a given year, no new entrants are added to the life table for that year.

2. The ratio of behavioral Ph.D.'s to non-behavioral Ph.D.'s employed in behavioral fields remains constant. That is, both Ph.D. populations increase or decrease at the same rate. The same assumption is made for the models of biomedical and non-biomedical Ph.D.'s in biomedical employment.

3. The age/destination state proportionate distribution, D_x, is taken from the age/destination distribution observed among new entrants between 1985 and 1991.

4. The age/origin state distribution for the current population, $N_{x,1991}$, is calculated by taking the age-specific origin state distribution among the current Ph.D. population between 1985 and 1991 and applying it to the age distribution of the 1991 current population.

5. The age-specific 2-year transition proportions, M_x, used to survive the current age-distribution is taken from the observed transition proportions between 1985 and 1991.

The committee's Panel on Estimation Procedures will extend this work and will prepare a separate report for release in 1994 on the role of multistate life table methods in the estimation of national need.

NOTES

1. Methodological detail is available on request from NRC/OSEP Studies and Surveys Unit (Memorandum by Peter Tiemeyer, September 30, 1993). A general discussion of multistate life tables can be found in Keyfitz (1985).

2. In our particular project, however, we began with transition rates as the basic input data, and derived other life table statistics from those rates.

3. We define R&D employment to be basic or applied research, management of R&D, or development and design of systems and products; it is based on the individual's self-report of primary or secondary work activity.

4. With respect to the treatment of missing data: in general, to enter into the calculation of a biennial transition table, an individual case was required to have valid survey data on age and Ph.D. field and valid data for both of the survey years (for that transition table) on employment field (biomedical, behavioral, etc.) and employment status (postdoctorate, employed, retired, etc.). We developed decision rules for the treatment of all of these variables to handle various conditions (available on request). For example, work activity (i.e., R&D vs. non-R&D) could be missing if the person's employment field was other than biomedical or behavioral (because one of the "states" of the model is "employed outside of Ph.D. field" and those who are out of labor force, retired, or out of the country could be missing employment field and "work activity".

5. The 1981-1991 average annual growth rates were: 4.25 percent per year for the biomedical sciences workforce and 3.5 percent per year for the behavioral sciences workforce.

6. Net separations are defined in this analysis as losses arising from death, retirement or outmobility to another state minus gains from inmobility of experienced scientists from other states of employment. Alternative definitions will be explored in subsequent work by the Panel on Estimation Procedures.

REFERENCES

Bowen, W.G. and J.A. Sosa
    1989   *Prospects for Faculty in the Arts and Sciences*. Princeton, NJ: Princeton University Press.

Keyfitz, N.
    1985   *Applied Mathematical Demography*. 2nd Ed. New York: Springer-Verlag.

National Research Council
    Forthcoming   *1991 SDR Methodological Report*. Washington, D.C.: National Academy Press.

Tiemeyer, P. and G. Ulmer
    1991   *MSLT: A Program for the Computation of Multistate Life Tables*. Center for Demography Working Paper 91-34, University of Wisconsin.

APPENDIX H

PROCEDURES USED TO ESTIMATE AWARDS, STIPENDS, AND COSTS

ASSUMPTIONS ABOUT PROGRAM SIZE

Table H-1 summarizes the distribution of awards by field and level and type of award for the fiscal years 1992-1999.[1] Increases are concentrated in the behavioral sciences, oral health, nursing, and health services research and in the Medical Scientist Training Program (MSTP). In addition, some postdoctoral awards in the clinical sciences have been reallocated to the predoctoral MSTP program.[2]

The number of fiscal year 1993 awards by field was estimated from statistics provided by NIH. The number of awards for FY 1994-1996 were estimated in several steps. The FY 1996 awards were first scaled to reflect the committee's recommended changes. These are summarized in Table H-2.[3] The awards for fiscal years 1997-1999 remain unchanged at FY 1996 levels, reflecting the committee's recommendations for stable award levels once their recommended award levels were reached.

STIPENDS

The committee also recommended steps to raise stipends to make them more competitive with other programs and to maintain the purchasing power of these stipends by adjusting them for inflation. Specifically, the committee recommended:

• raising the real value of stipends by FY 1996 to more competitive levels: approximately $12,000 per year in 1994 dollars for predoctoral awards and approximately $25,000 for new postdoctoral awardees with less than two years of research experience; and

• maintaining the real value of these stipends after FY 1996 through annual increases of three percent per year (the assumed annual rate of inflation).

Table H-3 summarizes the results of implementing the committee's recommendations with respect to stipend increases. The 1994 levels reflect NIH recommendations—an increase of $1,200 for the predoctoral awardees and $1,000 for the first two steps of the postdoctoral awardees. Using 1994 as a base and assuming an annual rate of inflation of 3 percent, the committee derived recommendations for 1996 stipend levels for predoctoral awardees ($12,731) and for postdoctoral awardees with less than 2 years of research experience ($26,523 and $27,623, respectively). The 1995 stipends were derived by linear interpolation between the 1994 and the 1996 levels. Stipends rise between 1994 and 1996 by about 12.5 percent per year for predoctoral awardees and by about 16.3 percent per year for postdoctoral awardees with less than one year of research experience to bring their levels to the targets set by the committee. The rate of increase moderates after 1996 and reflects only the three percent per year inflation increase recommended by the committee.

Because of the relatively low current level of stipends for postdoctoral awardees with less than two years experience, the stipend levels of postdoctoral awardees with two or more years of experience remained unchanged in fiscal years 1994-1995. But after fiscal 1995 these stipend levels had to be adjusted in order to prevent "inversion"—i.e., a situation in which the stipend of less-experienced postdoctoral awardees is greater than the stipend of more-experienced postdoctoral awardees. This was accomplished by raising the stipend level of all postdoctoral awardees by the same amount.

ESTIMATING INCREMENTAL STIPEND COSTS

The committee has developed incremental cost estimates for these recommendations. These incremental costs reflect only the recommended changes in the number of awards and the stipend levels.

153

APPENDIX H

TABLE H-1 Committee Estimates of Training Needs for Biomedical and Behavioral Research Personnel, by Field, Level and Type of Award, 1992 - 1999

Field and type of award	Year							
	1992	1993	1994	1995	1996	1997	1998	1999
Biomedical Sciences	8,687	9,007	9,007	9,007	9,007	9,007	9,007	9,007
Predoctoral awards	4,777	5,171	5,171	5,171	5,171	5,171	5,171	5,171
Fellows	290	360	360	360	360	360	360	360
Trainees	4,487	4,811	4,811	4,811	4,811	4,811	4,811	4,811
Postdoctoral awards	3,910	3,836	3,836	3,836	3,836	3,836	3,836	3,836
Fellows	1,550	1,533	1,533	1,533	1,533	1,533	1,533	1,533
Trainees	2,360	2,303	2,303	2,303	2,303	2,303	2,303	2,303
Behavioral	857	1,021	1,147	1,274	1,400	1,400	1,400	1,400
Predoctoral awards	534	672	748	824	900	900	900	900
Fellows	53	68	76	83	91	91	91	91
Trainees	481	604	672	741	809	809	809	809
Postdoctoral awards	323	349	399	450	500	500	500	500
Fellows	65	71	81	92	102	102	102	102
Trainees	258	278	318	358	398	398	398	398
Clinical (non MSTP)	3,012	2,974	2,975	2,918	2,860	2,860	2,860	2,860
Predoctoral awards	819	855	895	895	895	895	895	895
Fellows	19	29	20	20	20	20	20	20
Trainees	800	826	875	875	875	875	875	875
Postdoctoral awards	2,193	2,119	2,080	2,023	1,965	1,965	1,965	1,965
Fellows	64	68	80	110	160	160	160	160
Trainees	2,129	2,051	2,000	1,905	1,805	1,805	1,805	1,805
Clinical (MSTP)								
Predoctoral awards	806	822	889	955	1,022	1,022	1,022	1,022
Fellows	0	0	0	0	0	0	0	0
Trainees	806	822	889	955	1,022	1,022	1,022	1,022
Postdoctoral awards	0	0	0	0	0	0	0	0
Fellows	0	0	0	0	0	0	0	0
Trainees	0	0	0	0	0	0	0	0

TABLE H-1 (continued)

Field and type of award	Year							
	1992	1993	1994	1995	1996	1997	1998	1999
Nursing	257	236	340	420	500	500	500	500
Predoctoral awards	217	188	290	360	430	430	430	430
Fellows	142	112	195	240	285	285	285	285
Trainees	75	76	95	120	145	145	145	145
Postdoctoral awards	40	48	50	60	70	70	70	70
Fellows	12	12	15	20	20	20	20	20
Trainees	28	36	35	40	50	50	50	50
Oral Health	213	224	260	345	430	430	430	430
Predoctoral awards	77	97	125	210	290	290	290	290
Fellows	0	1	25	80	130	130	130	130
Trainees	77	96	100	130	160	160	160	160
Postdoctoral awards	136	127	135	135	140	140	140	140
Fellows	35	22	35	35	35	35	35	35
Trainees	101	105	100	100	105	105	105	105
Health Services	94	96	115	240	360	360	360	360
Predoctoral awards	35	30	55	180	300	300	300	300
Fellows	0	0	10	85	160	160	160	160
Trainees	35	30	45	95	140	140	140	140
Postdoctoral awards	59	66	60	60	60	60	60	60
Fellows	12	17	10	10	10	10	10	10
Trainees	47	49	50	50	50	50	50	50
Total	13,926	14,380	14,733	15,151	15,579	15,579	15,579	15,579
Predoctoral awards	7,265	7,835	8,173	8,595	9,008	9,008	9,008	9,008
Fellows	504	570	686	868	1,046	1,046	1,046	1,046
Trainees	6,761	7,265	7,487	7,727	7,962	7,962	7,962	7,962
Postdoctoral awards	6,661	6,545	6,560	6,555	6,571	6,571	6,571	6,571
Fellows	1,738	1,723	1,754	1,800	1,860	1,860	1,860	1,860
Trainees	4,923	4,822	4,806	4,756	4,711	4,711	4,711	4,711

TABLE H-2 Recommended Award Levels for FY 1996, by Field

Field	Number of awards
Biomedical	9,007
Behavioral	1,400
Clinical	3,994
non-MSTP	2,972
MSTP	1,022
Nursing	500
Oral Health	426
Health Services	360

Table H-4 summarizes the budgetary implications of these recommendations. The budget increases by $116 million between 1993 and 1999, a rise of 33 percent, or an annual rate of increase of 5.4 percent per year. The cost increase associated with the committee's recommendation to raise the number of awards is roughly one percent per year; the remainder, roughly 4.4 percent per year, is generated from the committee's recommendation to increase stipend levels to more competitive levels and to maintain this competitiveness through automatic annual cost of living increases.

The costs associated with the recommended increases in the number of awards and the stipend levels are also displayed in Table H-4. The estimated are displayed annually for the period 1993-1999 by type of award (i.e., predoctoral vs. postdoctoral, and fellowship vs. traineeship).

Columns (5) and (6) summarize the effects of increasing the number of awards on the training budget. These estimates hold average training costs constant at 1993 levels. Thus, the cost changes reflect only changes in the number of awards recommended. Total training costs rise by roughly $22 million, or 6 percent between 1993 and 1999. The average annual rate of increase is less than one percent. All of the increases are concentrated in the years 1994-1996, the period over which the committee's recommendations are implemented. Practically all of the cost increase occurs in predoctoral awards, reflecting the committee's recommendation for reallocation from postdoctoral to predoctoral awards.

The incremental costs associated with both the recommended increases in awards and stipends are displayed in columns (14) and (15). Incremental costs, summarized in columns (8)-(10), are derived by multiplying the recommended stipend increases by the number of awards affected by these increases. Recall that, for the years 1994 and 1995, the recommended stipend increases were awarded only to postdoctoral awardees with two or less years of research experience. The average fraction of postdoctoral awardees with two years or less in 1993 and 1994 was used to estimate the number of awards affected by the recommended stipend increase for fiscal year 1995.[4] For subsequent years it was assumed that all postdoctorates were affected. These incremental costs are cumulated in columns (11)-(13).

The costs of the recommended stipend increases are summarized in Table H-5. These increases will cost an estimated $94 million—roughly 26 percent of the 1993 training budget and 81 percent of the $116 million cost increment associated with all of the committee's recommendations. On an annual basis, these increases will average out to be roughly 4.4 percent of the 1993 budget, concentrated mainly in the years 1994-1996. The annual increases for the years 1997-1999 are roughly 2.5 percent per year.

NOTES

1. These numbers exclude NRSA support for research training through the MARC program and training in Primary Care Research.

2. In particular, 200 postdoctoral traineeships in the non-MSTP clinical sciences programs are reallocated to predoctoral traineeships in the MSTP program.

3. The scaling-up was accomplished by linear interpolation in the behavioral and MSTP programs. The rate of increase for the other fields reflected the Committee's judgment about the capability of the training system to absorb these increases. The increases by type of award within fields also reflects the Committee's wish to increase the relative importance of predoctoral awards in the clinical sciences and in oral health and health services research.

4. Individuals with more than two years of research experience constituted about 37 percent of the postdoctoral fellows and roughly 30 percent of the postdoctoral trainees in FY 1992.

TABLE H-3 Recommended Stipend Levels, 1993-1999

Year	Predoctoral Awards	Postdoctoral Awards by Number of Years of Research Experience							
		0	1	2	3	4	5	6	7
1993	$8,800	$18,600	$19,700	$25,600	$26,900	$28,200	$29,500	$30,800	$32,300
1994	$10,000	$19,600	$20,700	$25,600	$26,900	$28,200	$29,500	$30,800	$32,300
1995	$11,365	$23,061	$24,161	$25,600	$26,900	$28,200	$29,500	$30,800	$32,300
1996	$12,731	$26,523	$27,623	$29,062	$30,362	$31,662	$32,962	$34,262	$35,762
1997	$13,113	$27,318	$28,418	$29,857	$31,157	$32,457	$33,757	$35,057	$36,557
1998	$13,506	$28,138	$29,238	$30,677	$31,977	$33,277	$34,577	$35,877	$37,377
1999	$13,911	$28,982	$30,082	$31,521	$32,821	$34,121	$35,421	$36,721	$38,221

APPENDIX H

TABLE H-4 Estimated Cost of Awards and Stipend Increases, 1993-1999

Predoctoral programs

Year	Awards Fellow (1)	Awards Trainee (2)	Total cost (in thousands) (at 1993 stipend levels) Fellow (3)	Total cost (in thousands) (at 1993 stipend levels) Trainee (4)	Total cost (in thousands) (at 1993 stipend levels) Total (5)	Change in total cost (percent) (6)	Stipend increase (7)	Incremental cost of stipend increase (in thousands) Fellow (8)	Incremental cost of stipend increase (in thousands) Trainee (9)	Incremental cost of stipend increase (in thousands) Total (10)	Cumulative cost of stipend increase (in thousands) Fellow (11)	Cumulative cost of stipend increase (in thousands) Trainee (12)	Cumulative cost of stipend increase (in thousands) Total (13)	Total cost (in thous.) (14)	Percent change (15)
1993	570	7,265	9,908	141,115	151,023		0	0	0	0	0	0	0	151,023	
1994	685	7,487	11,907	145,427	157,334	4.18%	1,200	822	8,984	9,806	822	8,984	9,806	167,141	10.67%
1995	868	7,727	15,088	150,089	165,177	4.98%	1,365	1,185	10,547	11,732	2,007	19,532	21,539	186,715	11.71%
1996	1,046	7,962	18,182	154,654	172,835	4.64%	1,365	1,428	10,868	12,296	3,435	30,400	33,834	206,670	10.69%
1997	1,046	7,962	18,182	154,654	172,835	0.00%	382	400	3,041	3,441	3,834	33,441	37,276	210,111	1.67%
1998	1,046	7,962	18,182	154,654	172,835	0.00%	393	411	3,129	3,540	4,245	36,570	40,816	213,651	1.68%
1999	1,046	7,962	18,182	154,654	172,835	0.00%	405	424	3,225	3,648	4,669	39,795	44,464	217,299	1.71%
% change, 1993-1999					14.44%									43.88%	
avg annual % change					2.41%									7.31%	

Postdoctoral programs

Year	Awards Fellow (1)	Awards Trainee (2)	Total cost (in thousands) (at 1993 stipend levels) Fellow (3)	Total cost (in thousands) (at 1993 stipend levels) Trainee (4)	Total cost (in thousands) (at 1993 stipend levels) Total (5)	Change in total cost (percent) (6)	Stipend increase (7)	Incremental cost of stipend increase (in thousands) Fellow (8)	Incremental cost of stipend increase (in thousands) Trainee (9)	Incremental cost of stipend increase (in thousands) Total (10)	Cumulative cost of stipend increase (in thousands) Fellow (11)	Cumulative cost of stipend increase (in thousands) Trainee (12)	Cumulative cost of stipend increase (in thousands) Total (13)	Total cost (in thous.) (14)	Percent change (15)
1993	1,723	4,822	47,889	156,937	204,826		0	0	0	0	0	0	0	204,826	
1994	1,754	4,805	48,751	156,384	205,134	0.15%	1,000	649	1,778	2,427	649	1,778	2,427	207,561	1.34%
1995	1,800	4,755	50,029	154,756	204,785	-0.17%	3,461	2,305	6,089	8,394	2,954	7,867	10,821	215,606	3.88%
1996	1,860	4,710	51,697	153,292	204,989	0.10%	3,461	6,437	16,301	22,739	9,391	24,168	33,560	238,548	10.64%
1997	1,860	4,710	51,697	153,292	204,989	0.00%	796	1,481	3,749	5,230	10,872	27,917	38,789	243,778	2.19%
1998	1,860	4,710	51,697	153,292	204,989	0.00%	820	1,525	3,862	5,387	12,397	31,780	44,177	249,165	2.21%
1999	1,860	4,710	51,697	153,292	204,989	0.00%	844	1,570	3,975	5,545	13,967	35,755	49,722	254,710	2.23%
% change, 1993-1999					0.08%									24.35%	
avg annual % change					0.01%									4.06%	

Continued

TABLE H-4 (continued)
All programs

Year	Awards Fellow (1)	Awards Trainee (2)	Total cost (in thousands) (at 1993 stipend levels) Fellow (3)	Total cost (in thousands) (at 1993 stipend levels) Trainee (4)	Total cost (in thousands) (at 1993 stipend levels) Total (5)	Change in total cost (percent) (6)	Stipend increase (7)	Incremental cost of stipend increase (in thousands) Fellow (8)	Incremental cost of stipend increase (in thousands) Trainee (9)	Incremental cost of stipend increase (in thousands) Total (10)	Cumulative cost of stipend increase (in thousands) Fellow (11)	Cumulative cost of stipend increase (in thousands) Trainee (12)	Cumulative cost of stipend increase (in thousands) Total (13)	Total cost (in thous.) (14)	Percent change (15)
1993	2,293	12,087	57,797	298,052	355,849			0	0	0	0	0	0	355,849	
1994	2,439	12,292	60,657	301,811	362,468	1.86%		1,471	10,762	12,233	1,471	10,762	12,233	374,702	5.30%
1995	2,668	12,482	65,117	304,845	369,962	2.07%		3,490	16,636	20,126	4,961	27,399	32,360	402,322	7.37%
1996	2,906	12,672	69,878	307,946	377,824	2.13%		7,865	27,169	35,035	12,826	54,568	67,394	445,218	10.66%
1997	2,906	12,672	69,878	307,946	377,824	0.00%		1,880	6,791	8,671	14,706	61,359	76,065	453,889	1.95%
1998	2,906	12,672	69,878	307,946	377,824	0.00%		1,936	6,991	8,928	16,642	68,350	84,993	462,817	1.97%
1999	2,906	12,672	69,878	307,946	377,824	0.00%		1,993	7,200	9,193	18,636	75,550	94,186	472,010	1.99%
% change, 1993-1999					6.18%									32.64%	
avg annual % change					1.03%									5.44%	

TABLE H-5 Costs of Recommended Stipend Increases, 1994-1999

Year	Predoctoral	Postdoctoral	Total	Percent [a]
1994	$9,806	$2,427	$12,233	3.44%
1995	$11,732	$8,394	$20,126	5.66%
1996	$12,296	$22,739	$35,035	9.85%
1997	$3,441	$5,230	$8,671	2.44%
1998	$3,540	$5,387	$8,927	2.51%
1999	$3,648	$5,545	$9,193	2.58%
1994-1999	$44,463	$49,722	$94,185	26.47%

a As a percent of 1993 training budget.

APPENDIX I

BIOGRAPHICAL SKETCHES

COMMITTEE ON NATIONAL NEEDS FOR BIOMEDICAL AND BEHAVIORAL RESEARCH PERSONNEL

IRA J. HIRSH, Ph.D., *(Co-Chair),* Mallinckrodt Distinguished Professor (Emeritus), Washington University, Director, Central Institute for the Deaf, St. Louis, Missouri. Trained in experimental psychology, Dr. Hirsh has worked with people with hearing disabilities and done research in hearing, auditory perception, communication, speech and language, and communication disorders. He has won numerous awards and honors, and is a member of the National Academy of Sciences.

JOHN D. STOBO, M.D., *(Co-Chair),* Professor of Medicine, Director and Physician-in-Chief, The Johns Hopkins University and The Johns Hopkins Hospital, Baltimore, Maryland. Following studies in biological and biomedical sciences and a degree in medicine, Dr. Stobo's work focused on issues in rheumatology, internal medicine, and immunology. His research focuses on cellular immunology as well as the forces involved in the regulation of cell mediated and humoral immune responses. He is a member of the Institute of Medicine of the National Academy of Sciences.

HELEN BERMAN, Ph.D., Professor, Department of Chemistry, Rutgers University, Piscataway, New Jersey. Following training in chemistry, Dr. Berman has specialized in molecular biophysics and crystallography. Her research is in crystal structures of nucleic acid components and proteins as well as structural interactions between nucleic acids and proteins.

FRANCIS J. BULLOCK, Ph.D., formerly Senior Vice President, Research Operations, Schering-Plough Research Institute, Kenilworth, New Jersey and currently with the management consulting firm of Arthur D. Little, Inc.
Trained in pharmacy and chemistry, Dr. Bullock's interests are in molecular biology, immunology, and pharmacology. He has done industrial research in new drug discovery and in drug safety evaluation.

EDWIN C. CADMAN, M.D., Chair and Professor, Department of Internal Medicine, Yale School of Medicine, New Haven, Connecticut. Following training in biology and a degree in medicine, Dr. Cadman has specialized in oncology. His research has focused on biochemical pharmacology and cancer, including biochemical modulation and selective killing of cancer cells and transfer of drug resistance among malignant cells.

NANCY E. CANTOR, Ph.D., Chair, Department of Psychology, Princeton University, Princeton, New Jersey. Dr. Cantor has specialized in the fields of personality and social psychology as well as in personality and cognition.

ELI GINZBERG, Ph.D., Director, Eisenhower Center for the Conservation of Human Resources and A. Barton Hepburn Emeritus Professor of Economics, Graduate School of Business, Columbia University, New York, New York. Dr. Ginzberg is a senior member of the Institute of Medicine and the National Academy of Sciences and the author of 100 books in economics.

R. DUNCAN LUCE, Ph.D., Emeritus Victor S. Thomas Professor, Harvard University; Distinguished Professor and Director, Irvine Research Unit in Mathematical and Behavioral Sciences, University of California, Irvine, California. Trained in aeronautical engineering and mathematics, Dr. Luce has worked primarily in psychology, specializing in axiomatic theories of measurement, probabilistic models of choice and response times, psychophysics, and individual decision making. He is the author or co-author of 8 books and numerous scientific articles. In addition to many honors and awards, Dr. Luce is a member of the National Academy of Sciences.

RUTH McCORKLE, Ph.D., Professor, Adult Health and Illness Division, Director, Center for Advancing Care in Serious Illness, School of Nursing, and Associate Director, Division of Cancer Control, University of Pennsylvania Cancer Center, Philadelphia, Pennsylvania. Trained in nursing and mass communication, Dr. McCorkle has specialized in oncological nursing. Her major research interests include patient and family responses to illness and testing of nursing interventions on quality of life outcomes. She is the author of many publications, and is a member of the Institute of Medicine of the National Academy of Sciences.

RAYMOND S. NICKERSON, Ph.D., Senior Vice President (ret.), Bolt Beranek and Newman, Cambridge, Massachusetts. Trained in experimental psychology, Dr. Nickerson's professional responsibilities have included both research and management. His research interests are in basic and applied psychological research in several areas including perception, cognition, human factors, and person-computer interaction. He is the author of numerous books and publications, and has received several awards and honors.

MARY J. OSBORN, Ph.D., Professor and Head, Department of Microbiology, University of Connecticut Health Center School of Medicine, Farmington, Connecticut. Trained in physiology and biochemistry, Dr. Osborn has specialized in biochemistry and molecular biology. Her research interests are in the field of biogenesis of bacterial membranes. She has received numerous honors and awards, and is a member of the National Academy of Sciences.

CECIL W. PAYTON, Ph.D., Associate Professor, Department of Microbiology and Biology, Morgan State University, Baltimore, Maryland. Dr. Payton was trained in microbiology. His research interests are in the fields of microbial physiology and biochemistry.

RICHARD R. RANNEY, D.D.S., Dean, Dental School, University of Maryland at Baltimore, Baltimore, Maryland. Following training in dentistry, Dr. Ranney's work has focused on research, clinical teaching, and program administration in periodontics and dental education. His research interests focus on etiology and pathogenesis of periodontal diseases, particularly in relation to microbiology, immunology, and other host defense mechanisms. He is the author of numerous publications, and has received many awards.

MICHAEL ROTHSCHILD, Ph.D., Dean, Division of Social Sciences, University of California at San Diego, La Jolla, California. Trained in mathematical economics, Dr. Rothschild's research interests include market organization under uncertainty and imperfect information; law and economics; and models of non-optimizing behavior. Dr. Rothschild is a member of the American Academy of Arts and Sciences.

DONALD STEINWACHS, Ph.D., Professor and Chair, Department of Health Policy and Management, The Johns Hopkins University, Baltimore, Maryland. Trained in public health administration, Dr. Steinwachs' research interests lie in primary medical care; effects of availability, access, continuity, organization, and financing on cost and quality; information systems; impact of hospital cost containment strategies; models for health resource allocation; and health manpower planning and evaluation.

RICHARD F. THOMPSON, Ph.D, Keck Professor of Psychology and Biological Sciences and Director of the Neural, Informational and Behavioral Sciences Program, University of Southern California, Los Angeles, California. Following training in psychology and psychobiology, Dr. Thompson carried out postdoctoral research in neurophysiology. His area of research and scholarly interest is the broad field of psychobiology with a focus on the neurobiological substrates of learning and memory. He is the author of numerous books and articles, and has received many awards and honors. He is a member of the National Academy of Sciences.

PANEL ON ESTIMATION PROCEDURES

MICHAEL ROTHSCHILD, Ph.D., *(Chair),* Dean, Division of Social Sciences, University of California, La Jolla, California. (See Committee biographical listing).

EUGENE HAMMEL, Ph.D., Professor Emeritus of Anthropology and Demography, Department of Demography, University of California, Berkeley, California. Dr. Hammel's research interests include quantitative analysis of anthropological data, computer microsimulation of demographic and social patterns, and analysis of historical demographic data. He is a member of the National Academy of Sciences.

ALAN KRUEGER, Ph.D., Professor, Woodrow Wilson School of Public and International Affairs, Princeton University, Princeton, New Jersey.

ROBERT MARE, Ph.D., Director, Center for Demography and Ecology; Professor, Department of Sociology, University of Wisconsin, Madison, Wisconsin.

AAGE SØRENSEN, Ph.D., Professor, Department of Sociology, Harvard University, Cambridge, Massachusetts. Trained in sociology, Dr. Sørensen's research interests include the study of labor markets and careers, school organization and education processes, and models and methods for the analysis of longitudinal data.